D0463335

Entering Research

A Facilitator's Manual

ENTERING RESEARCH
A Facilitator's Manual

Workshops for Students Beginning Research in Science

Janet Branchaw
University of Wisconsin–Madison

Christine Pfund
University of Wisconsin–Madison

Raelyn Rediske
University of Wisconsin–Madison

Part of the
W.H. Freeman Scientific Teaching Series

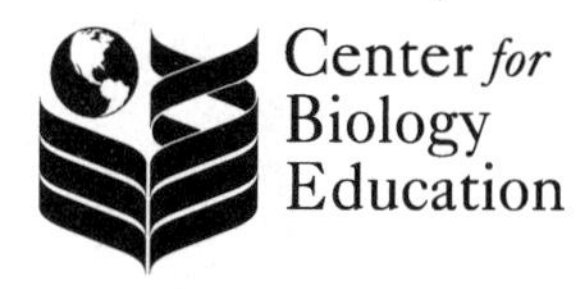

This material is based upon work supported by the National Science Foundation under Grant No. 0731564 awarded to Janet L. Branchaw. Any opinions, findings, and conclusions or recommendations expressed in this material are those of the author(s) and do not necessarily reflect the views of the National Science Foundation

Contact Information
Janet Branchaw
Phone: (608) 262-1182
Email: branchaw@wisc.edu

Executive Acquisitions Editor:	Susan Winslow
Project Manager:	Sara Ruth Blake
Associate Director of Marketing:	Debbie Clare
Cover and Text Designer:	Mark Ong, Side By Side Studios
Production Coordinator:	Ellen Cash
Composition:	Susan Riley, Side By Side Studios
Printing and Binding:	RR Donnelley

Scientific Teaching Series Editor:	William Wood, University of Colorado, Boulder
Scientific Teaching Series Editor:	Sarah Miller, HHMI Wisconsin Program for Scientific Teaching, University of Wisconsin–Madison

On the cover: corbis image 42-21876177 Collection Solus
© Jim Barber/Corbis

Supplementary materials are available online at:
www.whfreeman.com/facultylounge/majorsbio

Reprinted with permission of the American Society for Cell Biology (ascb.org), publisher of CBE–Life Sciences Education (www.lifescied.org) and a resource for educational tools, meetings, and more. CBE–Life Sciences Education is supported in part by a grant from the Howard Hughes Medical Institute.

Library of Congress Control Number: 2009941777

ISBN-13: 978-1-4292-5857-9
ISBN-10: 1-4292-5857-8

Printed in the United States of America

First printing

W.H. Freeman and Company
41 Madison Avenue
New York, NY 10010
Houndmills, Basingstoke
RG21 6XS, England
www.whfreeman.com

Contents

Foreword

Discovery is electrifying to the intellect and rewarding to the spirit. It nourishes the basic human drive of curiosity and promotes learning. Consequently, it is no surprise that participation in research impels many undergraduates to persist in science. Many factors, however, conspire to make undergraduate research experiences successful or not. Key among these is the guidance given to undergraduates about navigating the research context. Some students arrive in the research lab understanding lab culture with realistic expectations of the lab and themselves. Others have only a vague sense of what research is and what they are supposed to be doing. Some research lab groups hover over undergraduates; others provide great latitude that can veer into neglect.

Entering Research presents a solution to this variation by offering undergraduates guidance in the research arena. The book is a guide for a year-long course in research that prepares undergraduates for and then escorts them through the research experience. It helps them find research labs, teaches them the etiquette of approaching faculty, and orients them to the often alien culture of the lab. It provides them with a welcoming forum filled with other undergraduates sharing the same experience in which they can ask questions of their comrades that they may not feel comfortable asking in the research lab. They may be embarrassed to ask a mentor to explain a concept for the third time, but the *Entering Research* course offers them a context in which to ask for and receive an explanation from a peer. They may not know how to determine whether they are meeting their mentor's expectations. *Entering Research* helps them design a plan to find out the answer. Reducing the mystery of the research lab and equalizing the experience will enhance success of diverse students in the full spectrum of lab environments. In sum, *Entering Research* fosters more confident and thoughtful researchers who are more likely to satisfy their mentors and themselves in the research experience. This, in turn, will encourage more students to persist in science and more faculty to open their labs to undergraduate researchers.

Entering Research provides a stepwise guide to teaching a course about research. Developed over many years of teaching such a course, the content has been honed by instructors and their undergraduates and addresses every one of the critical issues raised by undergraduate researchers. The guide is complete, providing instructors with reading material, discussion points, leading questions, and assignments. Much like its prequel, *Entering Mentoring,* the guide makes it easy to teach with little preparation. The course is intended to be taught alone or in coordination with *Entering Mentoring*—its counterpart that is intended to train mentors. The content of the courses has been carefully designed to align the undergraduate and mentor experiences. Assignments are timed so that the experiences of the mentors and undergraduates mutually reinforce and make the instruction of each course easier.

Entering Research represents a substantive advance in undergraduate research. It provides the tools to fortify a new generation of scientists to navigate the shoals of research to find the joy and thrill of discovery in the lab. It will bring a more diverse group of students to successful research careers, enriching the scientific community and enhancing the creativity of the research that emerges from it.

Jo Handelsman
Professor of Molecular, Cellular and Developmental Biology, Yale University

Preface

The Bio 2010 report from the National Academies (NRC, 2003) recommends that "All students should be encouraged to pursue independent research as early as is practical in their education." *Entering Research* addresses this recommendation by outlining a framework to use with beginning undergraduate student researchers in the sciences. For students whose experience with science has been primarily in the classroom, it can be difficult to identify and contact potential mentors, and to navigate the transition to a one-on-one, mentor–student relationship. This is especially true for those who are new to research, or who belong to groups that are underrepresented in research. The *Entering Research* workshops offer a mechanism to structure the independent research experience, and help students overcome these challenges.

The goals of the workshops are to create a supportive learning environment to introduce students to the culture of research, to teach students valuable skills needed to become effective researchers, and to alleviate some of the work of faculty and lab personnel associated with mentoring novice student researchers. A safe learning community of peers is built through the sharing of experiences, and provides a place where students can begin to develop their identities as researchers. The student research experiences provide the content of the workshops, while an experienced researcher facilitates the workshops, acting as a guide on the side.

The name and format of the workshops were inspired by *Entering Mentoring*, a seminar to prepare scientists to mentor (Handelsman, Pfund, Miller Lauffer, and Pribbenow, 2005, *Entering Mentoring: A seminar to train a new generation of scientists.* Madison: University of Wisconsin Press). Offering *Entering Research* and *Entering Mentoring* in parallel should contribute to building strong mentor–student relationships, a critical component for successful research experiences (see appendix for parallel syllabi). In particular, the *Entering Research* workshops aim to empower students to take responsibility for their part of the mentor–student relationship, thus alleviating some of the time and energy normally required of mentors working with beginning researchers.

Though originally designed as a two-semester (one credit per semester) seminar series for undergraduates in the biological sciences engaging in their first, yearlong research experience, the materials in this manual can easily be adapted for a number of venues, with beginning researchers at any level and in any science discipline. For example, they can be used as individual, one-time workshops; as an intensive summer research program for undergraduates or pre-college students; as a series of professional development workshops for beginning graduate students; or as a way to support students working in an individual faculty member's research group. Samples of some of these adaptations are provided in the appendix.

Development of the materials was informed by several studies that have identified the benefits of undergraduate research (Kardash, 2000 *Journal of Educational Psychology*; Seymour, 2004 and 2006 *Science Education*; Lopatto, 2004 and 2007 *Cell Biology Education*, Russell, 2007 *Science*), by the authors' extensive experience working with beginning researchers and their mentors, and by the feedback and personal experiences of students and facilitators who participated in early offerings of the workshops. As a complement to these core materials, there is an on-line companion website, the Entering Research Faculty Lounge (www.whfreeman.com/facultylounge), where links to current resources and readings that address the workshop topics are available and frequently updated.

We strongly believe in the value of undergraduate research and have enjoyed developing and facilitating this workshop series. We hope that you and your students will likewise enjoy and benefit from it.

Janet Branchaw
Christine Pfund
Raelyn Rediske

Acknowledgments

This facilitator's manual is based on work supported by

- the National Science Foundation under grant No. 0731564 awarded to Drs. Janet Branchaw and Christine Pfund at the University of Wisconsin–Madison. Any opinions, findings, and conclusions or recommendations expressed in this material are of the authors and do not necessarily reflect the views of the National Science Foundation.
- the Howard Hughes Medical Institute under a Professors Program grant awarded to Dr. Jo Handelsman at the University of Wisconsin–Madison.
- the University of Wisconsin–Madison Institute for Cross-College Biology Education.
- the University of Wisconsin–Madison Center for Biology Education.

The authors offer special thanks to several faculty and professional staff members at the University of Wisconsin–Madison who facilitated early versions of the Entering Research workshop series. Their insightful suggestions and feedback contributed extensively to the development of this book. These individuals are David Abbott, Jennifer Ahrens-Gubbels, Nicholas Balster, Janet Batzli, Christopher Day, Erik Dent, Kevin Eliceiri, Michelle Harris, Jane Harris-Cramer, Liza Holeski, Brian Manske, Brian Parks, Thomas Sharkey, David Wassarman, and Christiane Wiese.

In addition we thank the students who participated in early versions of workshops. Their feedback was the key in refining the activities and assignments.

Finally, special thanks to our reviewers, Dr. Clarissa Dirks of Evergreen State College and Dr. Michelle Withers of West Virginia University, for their insightful reviews and helpful comments.

Overview

The materials in this manual are organized for use as a two-credit, two-semester workshop series (one credit per semester), Entering Research Parts 1 and 2 (see figure on page 2). However, they can easily be mixed and matched to create customized workshops.

Getting Started

Each workshop session is designed to last 1 hour and contains specific goals, an outline, facilitator notes, and student materials. The session activities are identified as core or optional: **core activities** directly address the session goals, and **optional activities** go beyond the main goals. Before each session, the facilitator should review the facilitator notes and select the activities to use.

Recruiting Students—Students who have not yet begun research should be recruited for Part I, and students who have been working on an independent research project with a mentor for at least 1 semester should be recruited for Part II. In addition to posting and distributing flyers to academic advisors, targeted invitations may also be sent to students enrolled in introductory science courses. Recruiting for the fall semester should begin at the end of the previous spring semester. Distribution of the pre-assignment to find a research mentor should occur as soon as students are signed up for the workshop, ideally at the end of the spring semester or at the beginning of summer. If the Entering Mentoring

Before	Entering Research Part 1	Entering Research Part 2	After
Facilitator: · design syllabus · set up electronic course management system (*optional*) · reserve meeting room · recruit students · recruit facilitators for additional sections (*optional*) · distribute pre-survey (one week prior to start) · distribute pre-assignment (two to three months prior to start) Students: · complete pre-survey · contact potential mentors	Facilitator: · facilitate sessions · weekly facilitator meeting (*optional*) Students: · identify a research mentor · establish common goals and expectations with research mentor · establish positive relationships with research group members · learn techniques · learn to document research · begin doing research · learn to read scientific literature · define research project in a proposal · engage in peer review	Facilitator: · facilitate sessions · weekly facilitator meeting (*optional*) Students: · reaffirm goals and expectations with research mentor · continue work on research project · discuss research ethics · explore research careers · learn to analyze and present data · contextualize research in society · present research findings · engage in peer review · write mini-grant proposing the next phase of the research project	Facilitator: · distribute post-survey (one week after end) · analyze survey data Students: · complete post-survey · continue next phase of research

seminar is being offered in the fall semester as well, then any mentor signed up for it, who is still looking for a student mentee, can be put in contact with the students looking for mentors.

Recruiting Facilitators (if offering multiple sections)—If there are more students interested in taking the workshop than can be readily accommodated in one section, then multiple sections may be offered. Experienced researchers, such as faculty, staff, and post-doctoral fellows, can be recruited as additional facilitators. In particular, those who are interested in developing their facilitation skills and committed to mentoring beginning researchers are excellent candidates.

It is tempting for those who know so much about research to teach the subject, rather than facilitate a discussion about it based on student experiences.

Regular facilitator meetings help to alleviate this temptation by providing a place where facilitators can share their experiences, brainstorm strategies to deal with challenges, and maintain an open dialog about how to be an effective facilitator. In addition, a list of common challenges in facilitating and suggested strategies for how to deal with them can be found in the appendix.

Workshop Goals and Student Learning Objectives

The primary goal of the Entering Research workshop series is to establish a learning community where beginning student researchers can get the most out of their research experience by sharing and learning from one another. As part of **Entering Research Part I**, students find a research mentor, write a research project proposal, and begin doing research. In **Entering Research Part II**, they complete their research, present their findings in a public venue, and write a mini-grant proposal for the next phase of their project. To achieve these goals, they work on learning objectives in three major areas:

1. **Research Process Skills**
   Students will
   - define a research question, design question, or hypothesis for their project.
   - find and evaluate relevant primary literature and background information related to their project.
   - design experiments to test their hypothesis.
   - learn the techniques needed to do their experiments.
   - learn and follow appropriate protocols for documenting their research.
   - analyze their experimental data.
   - use logic and evidence to build arguments and draw conclusions about their data.
   - define future research questions.

2. **Communication**
   Students will
   - explain the focus of their group's research, how individual research group members and projects are connected, and how the research in their group contributes new knowledge in their discipline.

- connect their research to issues relevant to society at large.
- effectively communicate their research findings in oral and written scientific formats.
- connect their research experience to what they have learned in courses.

3. Professional Development

Students will

- establish and maintain a positive relationship with their mentor by agreeing on common goals and expectations for the research experience, and revisit those goals and expectations regularly.
- define their roles and responsibilities as a member of their research group.
- define and contribute to discussions about the forms and consequences of scientific misconduct.
- contribute to peer review in the learning community and explain the role of peer review in science.
- know the mechanisms for funding research.
- identify research career options in their discipline.

Entering Research Part I

Dates	Topics	Assignments Due
Session 0 (Before start)	Finding a Research Mentor	**PRE-survey** · Identify potential mentors
Session 1	Introduction to the Workshop Series and Finding a Research Mentor	· Contact potential mentors
Session 2	The Nature of Science	· Research Experience Expectations
Session 3	Searching the Literature for Scientific Articles	· Research Topic and Key Words
Session 4	Reading Scientific Articles and Mentoring Styles	· Scientific Article Critique
Session 5	Your Research Group's Focus	· Your Research Group's Focus
Session 6	Establishing Goals and Expectations with Your Mentor	· Mentor Biography · Mentor-Mentee Contract
Session 7	Who's Who in Your Research Group	· Research Group Diagram
Session 8	Documenting Your Research	· Your Group's Research Documentation Protocol
Session 9	Defining Your Hypothesis or Research Question	· Visiting Peer Research Group · Background Information and Hypothesis or Research Question
Session 10	Designing Your Experiments	· Experimental Design & Potential Results with Timeline
Session 11	Research Proposal Review Draft #1	· Research Proposal Draft #1 · Peer Reviews
Session 12	Research Proposal Review Draft #2	· Research Proposal Draft #2 · Peer Reviews
Session 13	Final Research Proposal Presentations	· Final Research Proposal
Session 14	Final Research Proposal Presentations (continued)	· Research Experience Reflections **POST-Survey and Workshop Evaluation**

Entering Research Part II

Dates	Topics	Assignments Due
Session 15	Introduction to the Workshop Series and Science Communication	**PRE-survey**
Session 16	Research Project Outlines and Scientific Abstracts	· Research Project Outline & Science Abstract
Session 17	Research Project Outlines and Scientific Abstracts (continued)	· Reflecting on Your Mentoring Relationship
Session 18	Science and Society	· None
Session 19	Peer Review of General Public Abstracts	· Draft General Public Abstract · Peer Reviews
Session 20	Research Ethics	· Final General Public Abstract · Ethics Case Discussion with Mentor
Session 21	Making Effective Scientific Presentations	· Scientific Poster Hunt
Session 22	Research Careers	· Researching Research Careers
Session 23	Presentation Peer Review Draft #1	· Presentation Draft #1 · Peer Reviews
Session 24	Presentation Outside Review Draft #2	· Presentation Draft #2
Session 25	Final Presentations	· Final Presentation
Session 26	The Future of Your Project—Funding/Grants	· Research Group Funding
Session 27	Mini-Grant Proposal Peer Review	· Draft of Mini-Grant Proposal
Session 28	Research Experience Reflections and Celebration	· Final Mini-Grant Proposal · Research Experience Reflections **POST-Survey & Workshop Evaluation**

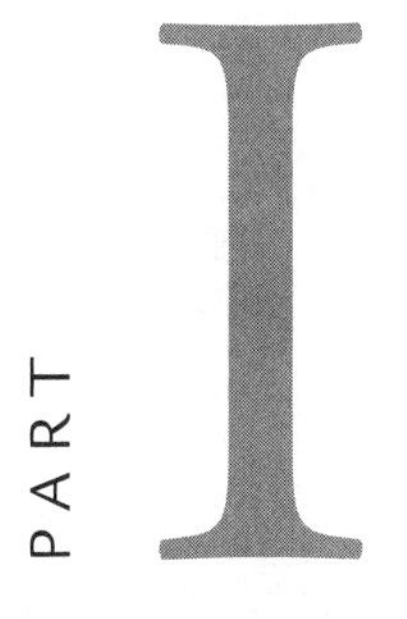

ENTERING RESEARCH
Session by Session

Facilitator Notes and Materials for Students

SESSION 0

Finding a Research Mentor

GOALS

Students will

- begin their search for a research mentor.

OUTLINE

Core Activities

Distribute the "Finding a Research Mentor" assignment

Monitor Student Progress

Introduce Mentors to the Entering Research Workshop Series

Optional Activities

Pre-survey (www.whfreeman.com/facultylounge)

Materials for Students

- Finding a Research Mentor Assignment

Possible Readings for Session 1

- WebGURU—**Guide to Research for Undergraduates**
  www.webguru.neu.edu/

Assignments for Session 1

- Finding a Research Mentor

FACILITATOR NOTES

Core Activities

Finding a Research Mentor

Distribute the "Finding a Research Mentor" assignment as soon as students sign up for the workshop series. This is best done electronically, either by electronic mail or through a workshop website.

A "mentor" is defined as the person with whom the student is working most closely on the research project. Students who attend large research universities may be assigned a graduate student, post-doctoral fellow, or scientist as a research mentor by the Principal Investigator, or professor, of a research group. If so, those students should consider this person their mentor when completing the assignments.

Monitoring Student Progress

Connect with students frequently to monitor their progress in finding a mentor before the workshop series begins. If the series will begin in the fall semester and the "Finding a Research Mentor" assignment is distributed at the end of the spring semester, it is helpful to send a monthly electronic reminder, asking students for an update on their progress in finding a mentor.

Introducing Mentors to the Entering Research Workshop

Once students have secured a research mentor, the facilitator should contact the mentor directly to introduce herself, and to let the mentor know what to expect from the Entering Research workshop series. In particular, there are several early assignments in which the student must meet with the mentor. Letting her know this before the series begins can help her prepare to work most effectively with the student. Here is a sample message to the mentor, and a list of materials to include with it:

Dear <Mentor>,

I am <Facilitator's Name> and your undergraduate research student, <Student's Name>, will be participating in my Entering Research workshop series. I'm writing to introduce myself and to tell you about the workshops.

The Entering Research workshop series is designed for undergraduate students who are beginning research in the sciences, and is taken concurrently with independent research credits. The main goals of Entering Research Part 1 (1 credit) are for students to establish a positive relationship with their mentor and define a research project. The main goals of Part 2 (1 credit, offered next semester) are to prepare a presentation for the campus-wide Undergraduate Symposium and to write a mini-grant proposal for the next phase of their research. The syllabus is attached for your review.

Please notice that there are three assignments on the syllabus that require the students to meet with their mentors: (1) Your Research Group's Focus, (2) the Mentor Biography, and (3) a Mentor–Mentee Contract. These assignments are designed to help students get to know you and your research group, and to help them establish a positive working relationship with you that is defined by shared goals and expectations. I've attached these assignments.

Please feel free to contact me if you have any questions about the workshop series or assignments.

Sincerely,
<Facilitator's Name>

Attachments:
- Syllabus
- Workshop Goals and Student Learning Objectives
- Your Research Group's Focus
- Mentor Biography
- Mentor–Mentee Contract

Optional Activities

Pre-Survey

If you are participating in the research study, ask students to complete the online survey, the link to which can be found on Freeman and Company's Faculty Lounge web site (www.whfreeman.com/facultylounge).

Wrap-Up

Assignments for next time:

- ▶ Finding a Research Mentor

Reflections and Notes

STUDENT MATERIALS

To download electronic versions of student materials go to Faculty Lounge (www.whfreeman.com/facultylounge)

Assignment
Due Session 1

Finding a Research Mentor

*Identify 5–10 potential research mentors and contact at least five of them **BEFORE the first day of class**. Use the attached pages to keep track of your progress and bring them to the first class meeting. Guidance about how to identify and contact potential mentors is provided below.*

Identifying Potential Research Mentors

1. Determine what most interests you in your discipline. In other words, define a research area (e.g. molecular biology, materials science, nanotechnology, plasma physics, analytical chemistry, computer architecture, etc.).
2. Do a search of campus websites to identify faculty working in your area of interest. Search through academic program listings, department web sites, student job sites, and undergraduate research databases if they are available. Talk to friends who are already doing research to get their advice about potential mentors. If you're not sure what research area interests you, then start by doing a general review of faculty research in the academic department in which you are majoring. But, don't be afraid to think broadly and explore research outside of your academic department, too!
3. Read the faculty research descriptions and generate a ranked list of potential mentors. Identify at least one thing about each person's research that is interesting to you and that you would like to know more about.

Contacting Potential Mentors

Email is a good way to make initial contact with potential mentors. By sending an email you give the mentor a chance to review your materials before responding. It is like the first step in an interview, so be sure it reflects your best effort (no spelling or grammatical errors!). If you are comfortable, it is also OK to phone or stop by a potential mentor's office to ask about a research experience.

Some things to consider when composing an email:

- Research mentors are very busy people, so keep it short and to the point (approximately 1 paragraph).
- Address the email using the mentor's official title (e.g. Professor, Dr.)
- Specifically refer to the mentor's research, and what you find interesting about it. Be sure to use your own words and not copy text from the research description on his or her web site.
- Be clear that you are looking for a research experience (vs. a dishwashing job) and what your main goal will be (e.g. shadowing someone in the lab to get exposed to research vs. doing an honors thesis research project).
- Highlight what you have to offer; what distinguishes you from other students (e.g. hard worker, experience, eager to learn, willing to stay more than one semester, persistent, specific courses you've completed that are relevant to the research).
- Show enthusiasm for learning how to do research!
- Finally, request that if the mentor is not able to take an undergraduate researcher, she recommend a colleague who might be able to.

Additional information you could include in an attached letter:

- Share that you are taking the Entering Research workshop series, and attach a copy of the syllabus.
- Give an estimate of the number of hours/credits you can be available to do research, and when you would like to begin, but leave room for negotiation.
- Give a *brief* overview of your academic credentials (e.g. GPA and relevant courses taken), or attach an electronic transcript.
- Provide your complete contact information (email, phone, mail).
- Show how you are different from other applicants.

Interviewing with Potential Mentors

- Be on time.
- Be yourself. But it will help if you come across as enthusiastic and motivated. Smile!
- Be ready to discuss why you want to do research in general (what are your academic and career goals?), and why you want to do research with this mentor specifically (what is it about his/her research that is interesting to you? Is there a particular project on which you would like to work?).

- Read about the research BEFORE you go to the interview. There is usually a research overview on the web with references/links to the group's published papers. Try to read one or two of these papers, and prepare some questions about them. Generally, mentors won't expect you to fully understand the research, but making the effort to learn about it on your own shows independence and motivation.
- Ask about the expectations of undergraduate researchers in the group (time commitment, credits, type of work). In general, three to five hours of research per week is worth one academic credit. However, this varies and you should ask how many hours the mentor expects per week per credit.
- Ask about who would be your direct mentor in the group (professor, postdoc, graduate student).
- Bring a copy of your transcript if you haven't already submitted one.

IMPORTANT

It can be challenging to connect with faculty research mentors, so ***be persistent, yet polite***. Ideally, give potential mentors a week to respond to your email before you follow up with a second.

Research groups have limited space, so it may be difficult to find a group that is looking for, or willing to take, another student. Do not take it personally if they decline your request. You may go through all ten (or more) potential mentors before you find a match. Stick with it! You will find someone.

Excellent Resource

WebGURU—Guide to Research for Undergraduates
http://www.webguru.neu.edu/index.php

1. Potential Mentor______________________________ Initial Contact Date:

Department: Follow-Up Contact Date:

Email: Phone Response:

Research Area: Interview Day & Time:

What I find interesting about the research:

2. Potential Mentor______________________________ Initial Contact Date:

Department: Follow-Up Contact Date:

Email: Phone Response:

Research Area: Interview Day & Time:

What I find interesting about the research:

3. Potential Mentor______________________________ Initial Contact Date:

Department: Follow-Up Contact Date:

Email: Phone Response:

Research Area: Interview Day & Time:

What I find interesting about the research:

4. Potential Mentor______________________________ Initial Contact Date:

Department: Follow-Up Contact Date:

Email: Phone Response:

Research Area: Interview Day & Time:

What I find interesting about the research:

5. Potential Mentor______________________________ Initial Contact Date:

Department: Follow-Up Contact Date:

Email: Phone Response:

Research Area: Interview Day & Time:

What I find interesting about the research:

6. Potential Mentor______________________________ Initial Contact Date:

Department: Follow-Up Contact Date:

Email: Phone Response:

Research Area: Interview Day & Time:

What I find interesting about the research:

7. Potential Mentor______________________________ Initial Contact Date:

Department: Follow-Up Contact Date:

Email: Phone Response:

Research Area: Interview Day & Time:

What I find interesting about the research:

8. Potential Mentor______________________________ Initial Contact Date:

Department: Follow-Up Contact Date:

Email: Phone Response:

Research Area: Interview Day & Time:

What I find interesting about the research:

9. Potential Mentor______________________________ Initial Contact Date:

Department: Follow-Up Contact Date:

Email: Phone Response:

Research Area: Interview Day & Time:

What I find interesting about the research:

10. Potential Mentor______________________________ Initial Contact Date:

Department: Follow-Up Contact Date:

Email: Phone Response:

Research Area: Interview Day & Time:

What I find interesting about the research:

SESSION 1

Introduction to the Workshop Series and Finding a Research Mentor

GOALS

Students will

- become familiar with the Entering Research Part I content, structure, and learning objectives.
- share their motivation for doing research, define specific goals for their research experience, and identify the contributions they expect to make to their research group.
- begin to form a learning community and establish ground rules for discussions.
- learn about their responsibility in contacting potential research mentors and discuss strategies to do so.

OUTLINE

Core Activities

Introductions

Entering Research Part I Overview

- Confidentiality and Discussion Ground Rules
- Syllabus
- Grading
- Assignments

Mentor Search Updates

Optional Activities

Research Experience Expectations

Materials for Students

- Name plates
- Syllabus, including student learning objectives
- Student Information Sheet
- Expectations of Research Experience Assignment
- Your Research Group's Focus Assignment
- Mentor Biography Assignment
- Summary of Expectations Discussion Assignment
- Reading(s) for Session 2

Possible Readings for Session 2

- Barker, K. (2004) Chapter 1—**General Lab Organization and Procedures**, *At the Bench, A Laboratory Navigator*, Cold Spring Harbor Laboratory Press.

- Slaughter, G.R. (2006) Chapter 1—**Research and Work Experience**, *Beyond the Beakers: Smart Advice for Entering Graduate Programs in the Sciences and Engineering*, Baylor College of Medicine, www.bcm.edu/gs/BeyondTheBeakers/Table%20of%20Contents.htm
- Webb, S. (6 July 2007) **The Importance of Undergraduate Research**, ScienceCareers.org.
- Lee, J.A., (1999) *The Scientific Endeavor*, **Chapter 1—A Primer on Scientific Principles and Practice**, Benjamin Cummings Press.
- Derry, G.N. (2002) Prologue—**What is Science**, *What Science Is and How It Works*, Princeton University Press.
- Derry, G.N. (2002) Chapter 1—**A Bird's Eye View: The Many Routes to Scientific Discovery**, *What Science Is and How It Works*, Princeton University Press.
- Williams, J. (2008) *The Scientist*, October. **What Makes Science "Science?"**

Assignments for Session 2

- Expectations of Research Experience
- Discussion Questions—These can be posted on an electronic chat board, or handed out to complete for the next meeting.
  - What is science?
  - What kinds of questions can science answer?
  - How are research and knowledge connected?

To Do for Session 2

- (*optional*) Invite experienced undergraduate researchers to meet with the class during Session 2.
- Contact the students' mentors, if you have not already done so, to share information about your Entering Research workshop series (see sample message in Session 0 Facilitator Notes).

FACILITATOR NOTES

Core Activities

Introductions

Building a strong, safe learning community should begin on the first day of class. Students may not have had the opportunity to participate in a small group workshop as an undergraduate, especially if they have been taking large introductory courses, so making sure that each student is introduced, welcomed to the group, and assured that his/her contributions will be valued is important. For students to be confident about and comfortable with learning from their peers (and not just the "expert" at the front of the room), they have to trust and be comfortable with one another and the facilitator. Though this trust will build as the course progresses, exercises like those described below can be used to help start the building process.

- Have each person pair up with someone he or she does not know, spend 2–3 minutes learning about their partner, and then introduce him/her to the group.
- Use name tents that sit on the table to help everyone learn names (these can be reused each week). Have students add four things that they want to share with the group to the corners of their name tent and present them when introducing themselves. For example: hometown, one or two words a parent would use to describe them, a guilty pleasure, their favorite undergraduate experience so far, favorite movie, favorite scientist, favorite musician/band, hobby, etc.
- Have students write something interesting about themselves on an index card (see list above), put the cards in a hat, and then pull cards from the hat at random (putting their own back should they happen to draw it). Students try to figure out whose card they have by asking yes/no questions, WITHOUT asking a direct question about what is on the card. Once they find their match, they visit with that person to learn more, and then introduce him/her to the large group.

Entering Research Workshop Series Overview

This workshop series is designed to complement and provide support for students beginning independent research in the sciences. **The main goals of Entering Research Part I are to help students find a research mentor, write a research project proposal, and begin doing research.** As part of the workshop series, students informally share their research experiences and learn about the diversity of experiences available through their peers. The "content" of the workshops comes from the students. Their engagement in the discussions is key, and the more they contribute, the more they will gain. The facilitator of the workshop provides a framework for the discussions, relevant background reading materials, and structured assignments. The assignments are designed to help students develop positive relationships with their research mentors, define themselves as a member of the research community, and understand and communicate their research. Advice from former undergraduate researchers to new researchers is presented in session 28 and may be shared with students at the 1st workshop.

Confidentiality and Discussion Ground Rules

Since the success of the workshops depends on creating an open, safe place for students to share their experiences, it is important to address confidentiality and respect for fellow group members on the first day of class. Students need to know that they can trust their peers to keep the experiences they share confidential (e.g. difficult interactions with their research mentors). Establishing ground rules for discussion is an effective way to address confidentiality and respect, and helps students learn how to effectively engage in group discussions. Ideally, the group will generate their own list of ground rules, but the facilitator may also generate a list of ground rules for the group to reflect on, discuss, and agree to. Some examples of ground rules include

- Everyone participates
- No dominating the discussion
- Do not interrupt
- Everything that is said here, stays here
- Respect differences in opinion
- No side conversations
- Listen attentively and carefully
- Criticize constructively
- Ideas should be separated from individuals—no personal attacks

In addition to establishing ground rules, individual behaviors can be addressed using the handout of constructive and destructive group behaviors included on Freeman and Company's Faculty Lounge web site (www.whfreeman.com/facultylounge). Students can be asked to reflect on their tendencies to engage in these behaviors and encouraged to work on self-monitoring those behaviors. One strategy to help them do this is to write their constructive and destructive behaviors on the back of their nameplate. This will remind them to engage in the constructive behaviors more frequently, and to limit their destructive behaviors.

Syllabus and Learning Objectives

Hand out and briefly review your syllabus and course learning objectives. A sample syllabus can be found in the appendix.

Grading

Discuss the grading scheme. Below is a guide that aligns with the leaning objectives presented in this manual. However, each instructor should design a grading scheme that aligns with the learning objectives outlined in his/her syllabus.

10% Attendance

10% Pre-Class Discussion Questions

10% In-Class Participation

25% Assignments

10% Hypothesis or Research Question

15% Experimental Design and Potential Results

20% Research Proposal

Assignments

▶ **Weekly Discussion Questions & Student Discussion Leaders**

If students will be facilitating discussion, then the student facilitator guidelines should be distributed at this time (appendix). The guidelines outline strategies students can use to design activities to lead a group discussion. If you will be using an on-line discussion tool, make sure students know how to use it at this time.

- **Mentor/Lab Assignments**

  Introduce the assignments that require students to connect with their research mentors. The first due date for these assignments is Session 5 (Mentor Biography). Therefore, if students do not have a mentor at the beginning of the workshop series, they still have five weeks to find one, with the help of the facilitator if necessary. Once a student has a mentor, these assignments should be completed as soon as possible. As a consequence, the assignments may be completed and submitted asynchronously by the students in the group. **Importantly, the facilitator should send a brief letter describing the workshop series, a syllabus, and packet of these assignments to the students' mentors as soon as they are identified, so they know what to expect.** The students' primary mentors may not be professors, but rather graduate students or post-doctoral fellows. If this is the case, then students should be advised to consider the person with whom they are working most closely to be their mentor for these assignments. The assignments include

  1. A paragraph describing what the student's research group studies and why it is important (Due Session 5)
  2. A short biography of the student's mentor (Due Session 6)
  3. A mentor–mentee contract (Due Session 6)

  Note: The second and third assignments complement assignments in *Entering Mentoring.*

- **Research Project Proposal—Final Assignment**

  Give an overview of the final research project proposal assignment. Tell students that it will include an introduction with relevant background information, a hypothesis or research question, and proposed experiments. Reassure them that the components will be developed as mini-assignments throughout the semester and compiled at the end of the semester as a complete proposal.

Research Mentor Updates and Course/Research Schedules

To monitor student progress in securing a mentor, ask them to either give you the name of their mentor (Student Information Sheet), or submit a copy of their potential mentor list, indicating those who they have already contacted.

To make sure that students have set aside time in their schedule to devote to research, ask them to share their course schedule with a peer. Often beginning research students do not realize the time they must commit to doing research, or the value of committing large blocks of time, rather than scattered hours, throughout the week. Sharing schedules opens the door for discussion.

Optional Activities

Research Experience Expectations Discussion

If time permits, you may ask students to complete their Research Expectations worksheets in class, rather than assigning them for the next class period. After completion, students can either share their answers in the large group, or break into small groups. The worksheet questions include

- Why do you want to do research?
- What specific goals do you hope to achieve in your research experience?
- What contributions will you bring to your research team?
- What is your greatest concern and what are you most excited about?

If you break into small groups, invite each small group to share one thing from its discussion with the large group, such as something they all had in common, or something they learned from their peers. Point out how the workshop will offer support to address concerns that are raised.

Wrap-Up

Invite students to share parting thoughts, concerns, or questions.

Assignments for next time:

- Expectations of Research Experience
- Discussion Questions:
  - What is science?
  - What kinds of questions can science answer?
  - How are research and knowledge connected?

Reflections and Notes

STUDENT MATERIALS

To download electronic versions of student materials go to Faculty Lounge (www.whfreeman.com/facultylounge)

Handout

Entering Research Part I: Student Information Sheet

	Student	Mentor #1 Name & Email (Professor or principal investigator)	Mentor #2 Name & Email (e.g. grad student, post-doc)
1.			
2.			
3.			
4.			
5.			
6.			
7.			
8.			
9.			
10.			
11.			
12.			

Assignment
Due Session 2

Expectations for Your Research Experience

1. Why do you want to do research?

2. What specific goals do you hope to achieve in your research experience?

3. What contributions will you bring to your research team?

4. What is your greatest concern, and what are you most excited about?

Assignment
Due Session 5

Your Research Group's Focus

Write one paragraph, in your own words, describing the focus of your group's research. Be sure to include the group's major research questions or hypotheses, the types of techniques they use to investigate these questions, and what area(s) of this work are most interesting to you.

To share this with the class, students will give "chalk talks" (informal oral presentations) of their research group's work in which each will be expected to draw a diagram on the board to explain the research.

Assignment
Due Session 6

Mentor Biography

Interview your mentor and write a short biography about him/her (2–3 paragraphs). Your mentor is the person with whom you are working most closely on your research project. This could be the professor in your research group, a graduate student, a post-doctoral fellow, or a scientist. In addition to the basic questions below, generate at least three additional questions to ask your mentor during the interview.

Basic Questions

1. Where did you grow up and what was it like there?

2. How and when did you decide to become a scientist?

3. Where did you do your training and how did you decide to attend those institutions? (undergraduate degree, graduate degree, etc.)

4. How did you decide on your disciplinary/research area? Have you done research in any other areas? If so, which?

5. What classes do you currently teach, or have you taught? Which was your favorite and why?

6. To which campus committees or organizations do you belong?

7. What hobbies do you enjoy?

Your Questions

1.

2.

3.

Assignment
Due Session 6

Mentor–Mentee Contract

Meet with your mentor to discuss what each of you expects from this research experience and complete a mentor–mentee contract. In the contract you will define a set of **common goals and expectations**. To prepare for this meeting, consider the topics listed below.

1. Why do you want to do research? Why does your mentor want to supervise an undergraduate researcher?

2. What are your, and your mentor's, career goals? How can this research experience and the mentor–mentee relationship help each of you achieve them?

3. What would success in this research experience look like to you? To your mentor?

4. How many hours per week and at what times/days do you and your mentor expect you to work?

5. Assuming a good fit, how long do you expect to work with this research group? Ideally, how long would your mentor like you to remain with the group?

6. What, if any, specific technical or communication skills do you expect to learn as part of the research experience? What specific skills would your mentor like you to learn?

7. Once you are trained in basic techniques, would you prefer to continue to work closely with others (e.g. on a team project), or independently? What level of independence does your mentor expect you to achieve, once basic techniques are learned? How will s/he know when you have reached this level?

8. How will you document your research results? Is there a specific protocol for keeping a laboratory notebook in your research group?

9. To whom do you expect to go if you have questions about your research project? Does your mentor expect you to come solely (or first) to him/her, or should you feel free to ask others in the research group? If others, can your mentor identify those in the group who would be good resource people for your project?

10. If you have previous research experience, what skills do you expect to bring to your new research group? If a student has previous research experience, is there anything the mentor should share about this research group that is unique and the student should be aware of?

Mentor–Mentee Contract

Mentee (print) ________________________ Mentor (print) ________________________

This contract outlines the parameters of our work together on this research project.

1. Our major goals are

   A. proposed research project goals –

   B. mentee's personal and/or professional goals –

   C. mentor's personal and/or professional goals –

2. Our shared vision of success in this research project is

3. We agree to work together on this project for at least ____ semesters.

4. The mentee will work at least _____ hours per week on the project during the academic year, and _____ hours per week in the summer.

   The mentee will propose his/her weekly schedule to the mentor by the _____ week of the semester.

   If the mentee must deviate from this schedule (e.g. to study for an upcoming exam), then s/he will communicate this to the mentor at least _____ (weeks/days/hours) before the change occurs.

5. On a daily basis, our primary means of communication will be through (circle)

   face-to-face/phone/email/instant messaging/______________

6. We will meet one-on-one to discuss our progress on the project and to reaffirm or revise our goals for at least _____ minutes _____ time(s) per month.

   It will be the (mentee's/mentor's) responsibility to schedule these meetings. (circle)

   In preparation for these meetings, the mentee will

   In preparation for these meetings, the mentor will

   At these meetings, the mentor will provide feedback on the mentee's performance and specific suggestions for how to improve or progress to the next level of responsibility through

   a. a written evaluation
   b. a verbal evaluation
   c. other ______________________________

7. When learning new techniques and procedures, the mentor will train the mentee using the following procedure(s) (e.g. write out directions, hands-on demonstration, verbally direct as mentee does procedure, etc.):

8. The proper procedure for documenting research results (laboratory notebook) in our research group is

9. If the mentee gets stuck while working on the project (e.g. has questions or needs help with a technique or data analysis) the procedure to follow will be

10. The standard operating procedures for working in our research group, which all group members must follow and the mentee agrees to follow, include (e.g. wash your own glassware, attend weekly lab meetings, reorder supplies when you use the last of something, etc.)

11. Other issues not addressed above that are important to our work together:

By signing below, we agree to these goals, expectations, and working parameters for this research project.

Mentee's signature ________________________________ Date ____________________

Mentor's signature ________________________________ Date ____________________

Professor's signature ________________________________ Date ____________________

SESSION 2

The Nature of Science

GOALS

Students will

- refine their definition of science and research.

OUTLINE

Core Activities

Check-in

Discussion of the Nature of Science

Activity: Science or Pseudoscience?

Optional Activities

Discussion with experienced undergraduate researchers

Materials for Students

- Activity: Science or Pseudoscience?

Possible Readings for Session 3

- None

Assignment for Session 3

- Identify research topic and associated key words

To Do for Session 3

- Arrange for students to meet at the library or a computer classroom for Session 3. Invite an expert (e.g. librarian) to teach the students how to use online databases to find scientific literature.

FACILITATOR NOTES

Core Activities

Check-in

Ask each student to give an update on his/her search for a research mentor.

Discussion of the Nature of Science

The goal of this discussion is to share ideas about the nature of science and research. What is science, and what is not science? If students are serving as facilitators, then they should have access to their peers' answers to the day's questions in time for planning their facilitation. Ideas that have emerged in response to these questions include

- **What is science?**
  - Science is using empirical methods to explain and understand things that go on around us.
  - Science is acquiring knowledge through methods like observation, problem solving, and experimentation.
  - Science is the discovery of the physical world through observation and experimentation.
- **What kinds of questions can science answer?**
  - Science can answer any question you can collect data on.
  - Science can answer anything that is testable.
  - Science cannot attempt to answer questions that require a value judgment.
- **How are research and knowledge connected?**
  - Research is usually done in order to gain knowledge.
  - Knowledge assists in research to gain new knowledge.
  - Research is a certain way of gathering knowledge.

Some additional questions to stimulate discussion:

- What does it mean to study the natural world?
- Who does science? Does it require a certain type of specialized training?
- Is there only one way to do science? Can you think of ways different from "the scientific method?"

Activity: Science or Pseudoscience?

Students are presented with a list of statements and asked to decide whether each is "scientific" or not, and to generate an argument to support their position. They should do this independently first, then gather in small groups to compare and discuss their conclusions.

Discussion questions:

- How do you decide whether something is scientific?
- Can you design experiments to test the statements you identify as scientific? Must one be able to test a statement in order to consider it scientific?

Optional Activity

Discussion with Experienced Undergraduate Researchers

Introduce the experienced undergraduate researchers and ask them to briefly describe their research and research group.

- How long they have been doing research?
- How did they find their research group/mentor?
- What is their "best piece of advice" about how to find a research mentor?

Wrap-Up

Assignments for next time:

- Identify research topic and associated key words
- Let students know where to meet for Session 3 (library or computer classroom)

Reflections and Notes

STUDENT MATERIALS

To download electronic versions of student materials go to Faculty Lounge (www.whfreeman.com/facultylounge)

Handout

Science or Pseudoscience?

Adapted from: www.indiana.edu/~ensiweb/lessons/conptt.html

1. Determine whether each of the following statements is scientific.
   A. Without sunlight (or comparable artificial light), green plants will die.
   B. If you are a "Taurus," your horoscope for today is "You are getting along better than ever with coworkers and other folks you may be forced to deal with every day. Your lucky numbers are 16 and 24."
   C. The earth began about 6,000 years ago and nothing will change that.
   D. Ships and planes passing through the Bermuda Triangle tend to sink and disappear.

2. What criteria did you use to determine whether each statement was scientific?

3. What is pseudoscience?

4. How can you tell the difference between science and pseudoscience?

SESSION 3

Searching the Literature for Scientific Articles

GOALS

Students will

- learn how to use on-line and library resources to search for scientific articles.

OUTLINE

Core Activities

Searching online databases

(*optional*) Field trip to the library

Optional Activities

Discuss strategies for reading scientific articles

Materials for Students

- Reading Scientific Articles handout
- Scientific Article worksheet
- Scientific article (generic article or article from mentor)

Possible Readings for Session 4

- Reading Scientific Papers handout
- Slaughter, G. R. (2006) Chapter 6—**Reading the Scientific Literature**, *Beyond the Beakers: Smart Advice for Entering Graduate Programs in the Sciences and Engineering*, Baylor College of Medicine. www.bcm.edu/gs/BeyondTheBeakers/Table%20of%20Contents.htm
- Schulte, B. A. (2003) **Scientific Writing and the Scientific Method, Parallel "Hourglass" Structure in Form and Content**, *The American Biology Teacher*, 65:8, pp. 591–594.
- Scientific articles that students get from their research mentors
- Generic scientific article: **A Guide to the Critical Reading of Scientific Papers**, www.greenwich.ac.uk/~bj61/talessi/tlr51.html
  - This site contains a bogus scientific article that can be read by the entire class ("The Effects of Grazing on Woodland Regeneration in the Highlands"), and an analysis of the article for the instructor.

Assignment for Session 4

- Scientific article worksheet. (Either a generic assigned article that the entire class reads, or, if all students have mentors, an article each selects with his/her mentor.)
- Discussion Questions:
  - What were the easiest and most challenging parts about reading the article?
  - Did reading an introductory textbook help you understand the article? Why or why not?
  - What other strategies did you use to "decipher" the article?
  - Do ideas from the scientific article link to your course work?

FACILITATOR NOTES

Core Activities

Searching Online Databases

During this session students learn how to do literature searches using online databases. They should arrive with the research topic they want to explore, and a list of key words associated with that topic. Most campuses have experts (e.g. librarians), who can be invited to class to teach students how to use the database tools. The session can either be held in a computer classroom, or at the library.

Optional Activities

Anatomy of a Scientific Article

The facilitator brings a sampling of scientific articles from different journals and different disciplines. Students are asked to look them over and to identify (1) the major sections (e.g. abstract, results), (2) what kind of information is presented in each of the sections, and (3) the differences between journals and disciplines.

Discussion on Strategies for Reading Scientific Articles

For next week, students will read a scientific article, which is a difficult task for most beginning researchers. Ask students to brainstorm strategies to use when reading their article. They can either do this first in pairs, or the group can brainstorm together. Some strategies that students have suggested include

- Read the abstract, introduction, and conclusion first to get an overview of the main points.
- Skip the methods section on the first read, since they may be confusing to one who has little experience doing experiments.
- Take notes as you read.
- Write out questions as you read.
- Search an introductory textbook or the web for terms or ideas that are difficult to understand. These sources often use simple language in their explanations.

Wrap-Up

Assignments for next time:

- Scientific article worksheet. (Either an assigned article that the entire class reads, or, if all students have mentors, an article each selects with his/her mentor.)
- Discussion Questions:
  - What were the easiest and most challenging parts about reading the article?
  - Did reading an introductory textbook help you understand the article? Why or why not?
  - What other strategies did you use to "decipher" the article?
  - Do ideas from the scientific article link to your course work?

Reflections and Notes

STUDENT MATERIALS

To download electronic versions of student materials go to Faculty Lounge (www.whfreeman.com/facultylounge)

Handout

Reading Scientific Papers

Adapted from materials developed by Nicholas Balster, PhD, Dept. of Soil Science UW–Madison, and Webguru: www.webguru.neu.edu/nuts_and_bolts/reading_the_technical_journal_article/

The BIG Picture

Before trying to understand the details presented in a scientific paper, it is wise to get an overview. Consider the following questions when first scanning papers.

- What central question/hypothesis are the authors asking/proposing?
- What assumptions are made both when proposing hypotheses and when evaluating them in light of the data collected?
- What data do they collect to assess their hypothesis?
- What is their conclusion given the data?

Basic Understanding

Scientific articles are typically organized in sections as outlined below. Knowing what types of information are present in each section allows one to more efficiently and effectively find information.

Title

Paper titles are usually succinct, stand-alone overviews of a paper's contents. Authors usually make an effort to include keywords that abstracting services could use in indexing the article. So, if you are new to a field and/or subject, it is useful to take note of the words used in the title as they may provide you with useful keywords to use in literature searches.

Abstract

The purpose of the abstract is to provide the reader with a succinct summary of the article. Thus, the abstract should provide information about the specific research problem being investigated, the methods used, the results obtained, and what the results of the study mean in the larger context of the research study and in some cases the field of study. This means that the abstract is a good place to look first if you are trying to decide whether or not the paper is relevant to your work.

Introduction

The introduction section generally provides an overview of the research problem being studied—why it is a worthy problem, what work has already been done by others to solve it, and what the authors may have already done in this area. Introductions are a good place to go if you are new to the subject.

- What is the main question the authors are interested in pursuing?
- What background research/pattern/theoretical prediction motivates this question?
- Why is this question interesting in light of the background they discuss?
- Do they offer one hypothesis or more than one?
- What assumptions are made when proposing the hypotheses?

Methods

The experimental section will provide detailed information on how the authors accomplished the experiments described in their paper. Such information typically includes sources for all reagents and/or materials used, names and models of all instrumentation used, methods for synthesizing any reagents, and provide quantitative information on the characterization of any new materials synthesized.

- Do their proposed methods critically test their hypotheses?
- Are any of their methods confounded?
- Did the authors use a creative method to evaluate their hypothesis?
- Are their methods simple and elegant or complicated and convoluted?
- Did they come up with a new technique to better evaluate a problem others have struggled with?

Results

Some articles will distinguish between "Results" and "Discussion" while others will combine this information into one section "Results and Discussion." In papers that contain two distinct sections ("Results" and "Discussion"), the data obtained from the study are introduced in the "Results" section and their interpretation is delayed until the "Discussion" section. In papers that contain one section ("Results and Discussion"), results are introduced and interpreted experiment-by-experiment.

- What do the data say about the hypotheses?
- Is there only one interpretation of the data?
- Are there any big surprises/unexpected results?

Discussion

Keep the following in mind:

- Do the authors say that the data support or reject the hypothesis?
- Do you agree with the authors' interpretation of the data?
- What novel insights are gained from the results?
- What do the results imply more generally for the field of interest? For other fields?
- What will the authors do next?

Sophisticated Understanding

With experience, reading the scientific literature in a given field will become easier. This includes the ability to better evaluate what is being presented, and the ability to ask more sophisticated questions.

- When reading papers be critical, but also pay attention to exciting findings, novel insights, and creative ideas. It's easy to criticize, but hard to praise!
- What critical experiment would you do to evaluate the proposed hypothesis?
- Form an opinion after looking at the data, *before* reading the author's interpretation and conclusions.
- Do you agree with the authors' interpretation or are there others?
- If more than one hypothesis is offered, is each exclusive, meaning that it proposes a distinctly alternative explanation that is incompatible with the others, or could some of the hypotheses be valid simultaneously?
- Are there compelling alternatives given the data?
- What assumptions are made about the effectiveness of the experiments or the accuracy of the data?

Assignment
Due Session 4

Scientific Article Worksheet

Title:

Authors:

Journal:

Volume, year, pages:

Scientific Article Overview

For each overview question indicate where in the article you found the information (e.g. title, abstract, introduction, etc.) Note that the answer may be found in more than one place.

1. Why is this research important?

2. What article would you read if you wanted to know more about this topic?

3. What is the hypothesis or goal of the research?

4. What are the major methods used?

5. What are the major results?

6. What conclusion did they make about their hypothesis/goal?

7. What new hypothesis/goal emerges from the results?

Scientific Article Critique

Use the questions below to critique the article.

1. Is the paper well written? How do you know?

2. Do the conclusions seem logical given the data presented? Why or why not?

3. Why are the conclusions important?

4. What were the best aspects of the research presented, and how could they be improved?

Additional Resources

1. What are the basic concepts that you need to understand in order to follow the science presented in this paper?

2. Identify a chapter/section in a textbook that outlines these basic concepts. Can you understand this material? Is reading this helpful to your understanding of the paper?

3. What other information or resources would help you better understand the paper?

Further Questions

Write at least five comments or questions about the article to discuss with your mentor.

1.

2.

3.

4.

5.

SESSION 4

Reading Scientific Articles and Mentoring Styles

GOALS

Students will

- be able to identify effective strategies for reading scientific articles.
- explore the pros and cons of different mentoring styles.

OUTLINE

Core Activities

Check-in

Discussion of reading scientific articles

Activity: Three Mentors

Optional Activities

Still need a mentor?

Materials for Students

- None

Possible Readings for Session 5

- None

Assignment for Session 5

- My research group's focus (student materials Session 1)

FACILITATOR NOTES

Core Activities

Check-in

Ask each student to give a brief update on his/her search for a mentor. **NOTE: All students should have a research group and mentor by now.**

Discussion of Reading Scientific Articles

The goal of this discussion is to share experiences of reading a scientific article and best approaches to reading articles. If the same article was assigned to all students, the discussion should focus on that article. If students read different articles that they selected with their research mentors, the discussion can focus on identifying the common challenges they faced and the strategies they used to overcome those challenges. Ideas that have emerged in response to these questions include

- **What were the easiest and most challenging parts of reading your article?**
  - Vocabulary (esp. common jargon to the field)
  - Background information
  - Understanding the importance of the study
- **What strategies did you use to "decipher" the article?**
  - Internet searches
  - Stopping to re-word or paraphrase difficult sentences
  - Starting at the discussion section then reading backwards
  - Talking with my mentor and using figures to help visualize
  - Highlighting, keeping notes, or making an outline
- **Did reading the introductory textbook help you understand the article? Why or why not?**
  - The textbook helped with some vocabulary, but the research was too specific for the book.
  - Could not help with the methods section.
  - My mentor actually assigned me some chapters and they helped me understand the basics of the lab's work.

- **Do ideas from the scientific article link to your course work?**
  - The article was very specific and my course work is more general.

Additional questions to stimulate discussion:

- How is the style of writing in scientific articles different from that in textbooks? How is it different from science articles written for the general public (e.g. *Scientific American*, *Discover* magazine)?
- Is writing an important skill for scientists to master? Why or why not?

Activity: Three Professors

This activity helps students develop a positive relationship with their mentor by asking them to reflect on different mentoring styles. They should complete the Three Professors worksheet independently, then share their answers with the group for discussion.

Discussion questions:

- What is the most important element in the mentor–mentee relationship for you now?
- Do you anticipate that this will change in the future? If so, how?
- What can you do to ensure that you get what you need out of this relationship with your mentor?

Optional Activities

Still need a mentor?

If any students are still searching for a research group to join, take time for the group to help this student identify how s/he might do things differently in the search.

Wrap-Up

Assignments for next time:

- My Research Group's Focus (student materials Session 1)

Reflections and Notes

STUDENT MATERIALS

To download electronic versions of student materials go to Faculty Lounge (www.whfreeman.com/facultylounge)

Activity

Three Professors

Read the discussions of three different types of professor as mentors and describe the advantages and disadvantages of working with each. Outline strategies you would use to overcome the disadvantages. After completing this exercise, outline the characteristics of your ideal mentor.

Professor 1

This mentor is very hands-on and likes to be the primary mentor for her undergraduate researchers. She works directly with students much of the time and wants to know everything that goes on all the time. She sets up weekly individual meetings and engages in frequent dialogue about the research.

Advantages:

Disadvantages:

Professor 2

This mentor is very famous and travels a great deal. Because of this, the researchers he advises have a formal system in which the senior researchers in the group act as mentors for the newer students. He keeps up to date on the progress of each student via frequent emails and meetings when in town.

Advantages:

Disadvantages:

Professor 3

This mentor is hands-off. She is around but likes to give students space to see how they handle independence. She typically has senior students informally mentor the newer ones. This mentor meets with each student once a month and holds regular structured lab meetings.

Advantages:

Disadvantages:

Describe *YOUR* ideal mentor

SESSION 5

Your Research Group's Focus

GOALS

Students will

- be able to clearly and concisely explain their group's research focus.

OUTLINE

Core Activities

Check-in

Presentations: What does your research group study?

Optional Activities

None

Materials for Students

- None

Possible Readings for Session 6

Slaughter, G.R. (2006) Chapter 4—**Making the Most of Mentor Relationships**, *Beyond the Beakers: Smart Advice for Entering Graduate Programs in the Sciences and Engineering*, Baylor College of Medicine. www.bcm.edu/gs/BeyondTheBeakers/Table%20of%20Contents.htm

Assignment for Session 6

- Mentor Biography (student materials Session 1)
- Expectations Discussion and Mentor–Mentee Contract (student materials Session 1)
- Discussion Questions:
  - How well did your goals and expectations for your research experience align with your mentor's?

- How does your project fit with the other projects your research group is studying? The broader area of your group's research?
- Was it difficult to generate a contract with your mentor? Why or why not?

FACILITATOR NOTES

Core Activities

Check-in

Ask each student to rate on a scale of 1–10 how difficult it was to write about their research group's research in their own words.

What Does Your Research Group Study?

Students distribute (or post on-line) their paragraphs about their research group's focus, and give a "chalk talk" (an informal oral presentation) to the group. Require them to draw on the board to clarify the information they are presenting, including definitions of "scientific lingo," and to get the audience engaged. To encourage discussion, you can require each student to ask at least two questions (overall) during their peers' presentations. In addition, you may ask the students to create a "web of projects" in which each student must make a connection between her research and the previous student's research.

A variation on this exercise is to pair students up to discuss their research, then have them describe and answer questions about their partner's research instead of their own. Though more challenging, it would be appropriate to do this with students who have been engaged in research for a while, rather than just starting out.

Optional Activities

None

Wrap-Up

Assignments for next time:

- Mentor Biography (student materials Session 1)

- Expectations Discussion and Mentor–Mentee Contract (student materials Session 1)
- Discussion Questions:
  - How well did your goals and expectations for your research experience align with your mentor's?
  - How does your project fit with the other projects your research group is studying? The broader area of your group's research? Why or why not?
  - Was it difficult to generate a contract with your mentor?

Reflections and Notes

STUDENT MATERIALS

To download electronic versions of student materials go to Faculty Lounge (www.whfreeman.com/facultylounge)

SESSION 6

Establishing Goals and Expectations with Your Mentor

GOALS

Students will

- write a short biography of their mentor to get to know him/her as a person.
- develop a mentor–mentee contract for their research experience.
- create an outline of the things they would like their mentor to be able to say about them in a letter of recommendation.

OUTLINE

Core Activities

Check-in

Discussion and sharing of mentor–mentee contracts

Optional Activities

Activity: Letter of Recommendation

Materials for Students

- Letter of Recommendation worksheet

Possible Readings for Session 7

None

Assignment for Session 7

- *Optional:* Share the letter of recommendation outline with mentor.
- Research Group Diagram
- Discussion Questions:
  - What do you like best about your research project and research group?
  - What do you find most challenging about your research project and research group?
  - To whom would you go if you had a problem with your immediate mentor?

FACILITATOR NOTES

Core Activities

Check-in

Ask all students to share the most interesting thing they learned about their mentor during the interview.

Discussion and Sharing of Mentor–Mentee Contract

The goals of this discussion are to identify reasonable goals and expectations for an undergraduate research experience, and to share how easy or difficult it was for students to have this conversation with their mentors. Ideas that have emerged in response to these questions include

- ▶ **How well did your goals and expectations for your research experience align with your mentor's?**
  - The student is still unsure what exactly he would like to get out of the research experience, but was happy with what the mentor defined.
  - The research experience is going to be much more independent than the student expected and her concern is whether there will be enough guidance in the beginning to get started.
  - Though the goals matched, they were focused on teaching and learning experimental techniques at this early stage in project development.
- ▶ **How does your project fit with the other projects your research group is doing—the broader area of research your group is engaged in?**
  - My project is vital to only a certain part of the lab. The lab is broken up to multiple projects and my project is localized to my mentor's group.
  - There are two projects going on in the group and mine is connected to only one of them.
  - My project is considered a "side project."
- ▶ **Was it difficult to generate a contract with your mentor?**
  - The goals and expectations discussion helped.
  - We still have a few things to settle, but got most of it done.

Additional questions to stimulate discussion:

- ▶ What were the most and least comfortable topics to discuss with your mentor?
- ▶ What are you most excited about, and what are you most concerned about after talking with your mentor?

Ask students to share the contracts they developed with their mentors. Encourage students to revisit their contracts with their mentor if there are items they did not discuss the first time. Some goals and expectations that students have outlined in the past include

Student's Expectations of Mentors

I expect my mentor to

- meet with me at least every few weeks.
- be open to my questions and to take time to think about them carefully.
- be patient with me because I am new to research.
- initially be directive, but eventually let me design and do experiments on my own.
- challenge and encourage me.
- teach me basic research techniques/procedures and safety protocols.
- help me define a project that is doable, yet relevant, and that keeps me busy.
- help me understand the basic scientific concepts and study design underlying my project.
- understand when I need to take time away from research to focus on my course work, and allow me to take it.
- listen to and consider my ideas seriously.

Mentor's Expectations of Student

I expect my student to

- be present and punctual when we have scheduled meeting times.
- work hard and give his/her best effort.
- manage his/her time efficiently and effectively when doing research.
- keep up with course work, but to let me know if he needs a break from research to focus on courses.
- make every effort on her own to understand the research our group does, but to ask questions when she does not understand.

- listen carefully, take notes, and follow instructions when being taught new techniques.
- follow all safety procedures.
- gradually gain independence, but to regularly communicate with me about what he is doing.
- be able to analyze her experimental data, generate logical conclusions based on that analysis, and propose future experiments, with assistance.
- work cooperatively, collaboratively, and respectfully with the other members of the research team.
- be attentive, creative, and contribute at research group meetings.

Optional Activities

Activity: Letter of Recommendation

Have students complete the Letter of Recommendation worksheets on their own first, then pair with another student to share ideas. Bring the large group together to discuss them. If students are comfortable with their mentors, they may take these letters to them to initiate a discussion about whether meeting the goals and expectations they agreed on will allow the mentor to address these things in a letter.

Wrap-Up

Assignments for next time:

- *Optional:* Share the letter of recommendation outline with mentor.
- Research Group Diagram
- Discussion Questions:
  - What do you like best about your research project and research group?
  - What do you find most challenging about your research project and research group?
  - To whom would you go if you had a problem with your immediate mentor?

Reflections and Notes

STUDENT MATERIALS

To download electronic versions of student materials go to Faculty Lounge (www.whfreeman.com/facultylounge)

Activity

Letter of Recommendation

One of the benefits of doing research is that you will get to know your mentor well and s/he will be able to write a detailed letter of recommendation on your behalf when you move to the next stage of your academic or professional career. As you begin the research experience, reflect on the expectations and goals you established with your mentor. Consider what you would like him/her to be able to say about you at the end of the research experience and complete this draft letter of recommendation.

Date

Dear Selection Committee,

I am writing this letter in support of [*your name*], who is applying for [*job of your dreams*]. I believe [*your name*] is an excellent candidate for this position because

1.

2.

3.

Over the past few years, [*your name*] has worked in my research group in the Department of [*your department*] at the [*your campus*]. [*Your name*] is very skilled in the following areas:

1.

2.

3.

In short, I believe that [*your name*] would be a wonderful asset to your department/program/unit. I strongly recommend him/her.

Sincerely,

[*Your mentor*]

Do the things outlined in this letter of recommendation align with the goals and expectations you established with your mentor? If not, how can you adjust your goals and expectations so that you will have the opportunity to engage in activities that allow your mentor to comment on these things?

Assignment
Due Session 7

Research Group Diagram

Draw a diagram to identify the people and projects in your research group. The diagram should represent how the projects are connected to one another, how the people are connected to one another, and how the projects and people are connected. The research group's overall area of study should be represented, and ideally encompass all parts of the diagram. Specifically include how you and your project fit in, and with whom in the group you see yourself collaborating.

SESSION 7

Who's Who in Your Research Group

GOALS

Students will

- become aware of the different sizes of research groups and the varied roles individuals play in research groups.
- be able to explain how their research project complements other projects in their research group, and how they will collaborate with the group members working on those projects.

OUTLINE

Core Activities

Check-in

Discussion of Research Group Structure

Presentations of Research Group Diagrams

Optional Activities

Sticky Situations

Case Study: Frustrated

Case Study: Overwhelmed

Materials for Students

- Case Study: Frustrated
- Case Study: Overwhelmed
- Reading on Documenting Research
- Assignment: Visiting Peer Research Groups

Possible Readings for Session 8

- Barker, K. (1998) Chapter 5—**Laboratory Notebooks**, *At the Bench, A Laboratory Navigator*, Cold Spring Harbor Laboratory Press.
- Slaughter, G. R. (2006) Chapter 5—**Data Management**, *Beyond the Beakers: Smart Advice for Entering Graduate Programs in the Sciences and Engineering*, Baylor College of Medicine. www.bcm.edu/gs/BeyondTheBeakers/Table %20of%20Contents.htm

Assignment for Session 8

- Reading on documenting research
- Your group's Research Documentation Protocol
- Bring your laboratory notebook (or a couple of copied pages from it) to class next week—or the facilitator can bring some examples.
- Discussion Questions:
  - Why is it important to accurately document your research, or keep a good lab notebook?
  - What are the key features of the notebooks in your research group? How were you instructed to document your research?

FACILITATOR NOTES

Core Activities

Check-in

Ask students to share whether they revisited their expectations and goals with their mentors after the letter of recommendation activity. Did anyone show his/her mentor the draft of the letter of recommendation?

Discussion of Research Group Structure

The goal of this discussion is to explore the variability in research group structures and research approaches. In particular, the discussion should highlight that research experiences depend a great deal on the research group, the people in it, and how it is structured, not just the research project. Ideas that have emerged in response to these questions include

- **What do you like best about your research project and research group?**
  - I am very excited about the research my group does and my project.
  - I feel very important and useful with all of the training that I am to undergo before work even begins.
  - I like the freedom I have to do the project in the ways that I see necessary.
  - I like that my project is a long term-one—I wanted to do research this summer, so this research provides me with a summer job.
- **What do you find most challenging about your research project and research group?**
  - The content is a little scary to me. Learning the jargon and protocol for this project is going to be very challenging, but I see it being worth it once I'm able to be very independent.
  - I think the most challenging part about being in my research group is going to be living up to the accomplishments of the other undergraduates that have been in the lab a few years.
  - The most challenging part is that no one checks my work so I spend a lot of time going over things so that my entries don't invalidate the research.

- My mentor is very quiet and English is not her native language. I'm worried that it will be difficult to communicate with her and that it will make my project harder to understand.

▶ **To whom would you go if you had a problem with your immediate mentor?**
- Other graduate students in the research group or the principal investigator.
- I would ask other members of my lab who have been there for a while to give me advice if I had problems with my immediate mentor. I would ask them what they thought of the problem, and ask for their advice on how to solve it.
- I don't really know whom I would go to about my mentor. I guess really the only thing would be to directly talk to him about any problems I might be having.

Additional questions to stimulate discussion:

▶ Does your research group have regularly scheduled group meetings? If so, how often and what is discussed? Are you expected to attend these meetings?

▶ How do members of the research group interact and communicate with one another?

Research Group Diagram Presentations

Students should compare their research group diagrams. Dividing them into smaller groups of three or four works well. Have each student describe his/her diagram and explain the relationships among the members represented in it. Discussion questions to consider:

▶ Does the diagram show a hierarchical structure? Where is the principal investigator or professor on the diagram (at the top, in the middle)? Where are the undergraduate students? Does the diagram reflect your access to the principal investigator or professor?

▶ How are the diagrams in each group similar? How are they different?

▶ Does the diagram show how each person and project in the group is related to the mentee's project?

Optional Activities

Sticky Situations

Pass out the list of sticky situations and ask students to individually write down what they would do in each situation. Bring the group together to discuss each situation and ask them to reflect on their research group diagrams during this discussion. How do the lines of communication defined on those diagrams help to define the best course of action in each situation?

Case Study: Frustrated

Pass out the case study and ask students to individually write down answers to the questions. Bring the group together to discuss the case.

Case Study: Overwhelmed

Pass out the case study and ask students to individually write down answers to the questions. Bring the group together to discuss the case.

Wrap-Up

Assignments for next time:

- Reading on documenting research
- Your research group's Documentation Protocol
- Bring your laboratory notebook (or a couple of copied pages from it) to class next week—or the facilitator can bring some examples.
- Discussion Questions:
    - Why is it important to accurately document your research, or keep a good research notebook?
    - What are the key features of the notebooks in your research group? How were you instructed to document your research?

Reflections and Notes

STUDENT MATERIALS

To download electronic versions of student materials go to Faculty Lounge (www.whfreeman.com/facultylounge)

Activity

Sticky Situations

1. Your mentor wants an experiment done this week, but you do not have the time because of exams. What do you do?

2. You have been working in the lab for 3 months and are not interested in the experiments. How would you approach your mentor?

3. Someone in your lab gives you a new protocol that they say is better than the one given to you by your mentor. Which protocol do you use?

4. Your mentor expects you to know everything and you are in over your head. The other graduate students in the lab are no help. What do you do?

Activity

Case Study: Frustrated

Jamal has been in his research group for almost three weeks and is disappointed with his project so far. When he interviewed with Professor Stanley, she described a molecular biology project that he would work on. However, his graduate student mentor, Roxanne, has not given him any molecular biology experiments, but instead tasks such as making media and growing bacteria. Other undergraduates in the lab seem to be doing things like cloning and sequencing genes. Jamal is getting frustrated, but doesn't want to complain or look ungrateful. What can he do?

1. To whom should Jamal go to discuss his frustration?

2. What strategies might he use to avoid appearing as though he is complaining?

3. How might having established specific goals and expectations with his mentor helped to avoid this situation?

Activity

Case Study: Overwhelmed

Ashley, a sophomore majoring in chemistry, has found an undergraduate research position at the Center for NanoTechnology. She started a couple of weeks ago and is excited about her research project, which involves working on the development of an automatic gene synthesizer, but she doesn't really understand it. She is a shy person and was completely overwhelmed at the first lab meeting. It was like nothing she had ever experienced and she understood very little of what was discussed. She won't take Introductory Biology until next year. At the meeting, she just nodded whenever they asked if she understood, because she didn't want to look stupid. Now she is terrified to talk to the scientists for fear that they will realize how little she really understands. Her mentor Sam, a biomedical engineering graduate student, is really nice, but also very busy. He told her to ask questions when she didn't understand something, but he is always engrossed in his work and she doesn't want to interrupt him. She has to write a one-page summary of her research project for the undergraduate research seminar class by the end of next week, and has no idea where to begin. What should she do?

1. Is there a way for Ashley to approach her mentor to ask questions that respects his busy schedule?

2. Who else beside her mentor could Ashley turn to for help?

3. What resources might she use to help herself better understand the research on her own?

Assignment
Due Session 8

Your Research Group's Documentation Protocol

Meet with your mentor to go over the protocol you must follow when documenting your research results. Outline that protocol and identify which parts of the process are common to your entire lab, and which are specific to your project. Where are the notebooks in your laboratory kept? Are they hard copy documents, or electronic files? Bring your lab notebook, or a copy of some pages from it, to our next class meeting.

Assignment
Due Session 9

Visiting Peer Research Groups

The goal of this assignment is to give you the opportunity to experience a diversity of research approaches and environments. Make arrangements with two of your peers in the workshop to visit their research groups. Write a short essay about your experiences, including a comparison of the three research groups. In particular comment on the following:

- How is the research space set up? Is it a laboratory, or some other kind of workspace?
- What kinds of interactions do you observe between research group members?
- How much diversity of activities and people is in the research group?
- Where is your mentor's workspace in relation to your classmates'? In relation to the PIs?
- Would you enjoy working in these other groups? Why or why not?

SESSION 8

Documenting Your Research

GOALS

Students will

- be able to explain why it is important to accurately document their research.
- be able to identify key elements in research documentation.

OUTLINE

Core Activities

Check-in

Discussion of Research Notebooks and Your Group's Research Documentation Protocol

Optional Activities

Can you decipher this?

Case Study: Keeping Data

Materials for Students

Case Study: Keeping Data

Possible Readings for Session 9

- Glass, D. J. and Hall, N. (8 August 2008) **A Brief History of the Hypothesis**, *Cell* 134: 378–81.

Assignment for Session 9

Background Information and Hypothesis or Research Question

FACILITATOR NOTES

Core Activities

Check-in

Ask each student to name one thing that should be included in each research notebook entry. Generate a comprehensive list.

Discussion of Laboratory Notebooks and Your Group's Research Documentation Protocol

The goals of this discussion are to identify the variability and commonalities in research notebook procedures, and to understand the importance of keeping an organized and complete notebook. Students should be able to describe how they were instructed by their research group to keep a notebook, and learn from their peers' other ways in which notebooks are kept. Ideas that have emerged in discussion of these questions in previous offerings include

- **Why is it important to keep a research notebook?**
  - To have clear documentation of experimental protocols and results for future reference.
  - The lab notebook is evidence of experimental findings.
  - The notebook belongs to the research group and supports the work they present and/or publish.
- **What are the key features in notebooks in your research group? (How were you instructed to organize it?)**
  - Date all entries.
  - Include a table of contents.
  - Some groups require electronic rather than paper notebooks (e.g. Google notebooks).
  - Write everything down in the notebook, including protocol details.
  - Organization and clarity are important.
  - Keep all notes and data in one place, rather than several different notebooks.
  - Write the objective and conclusion for each experiment.

Additional questions to stimulate discussion:

- Who owns the lab notebook? For whom is it written?
- How do your lab's research documentation protocols reflect the culture in your research group?
- How do they reflect the communication style in your research group?

Optional Activities

Can you decipher this?

Ask students to bring their laboratory notebooks (or copies of a couple of pages if it can not leave the lab) to class. Have them exchange notebooks and try to understand what the other person did in his experiment based on what is written in the notebook.

An alternative to this exercise is for the facilitator to bring in copies of pages from notebooks to be critiqued by the students.

Case Study Keeping Data

Pass out the case study and ask students to individually write down answers to the questions. Bring the group together to discuss the case.

Wrap-Up

Assignments for next time:

- Background Information and Hypothesis or Research Question

Reflections and Notes

STUDENT MATERIALS

To download electronic versions of student materials go to Faculty Lounge (www.whfreeman.com/facultylounge)

Activity

Case Study: Keeping Data

May, a junior, who has been doing research with Professor Jones for 2 years, is preparing to present her research results at the campus-wide Undergraduate Research Symposium. Because some of her findings are quite novel and contradict reports of similar experiments in the literature, Professor Jones asks to review the raw data before signing off on her presentation. When he reviews May's notebook, however, there are no hardcopy records of the data. Instead he finds the data on May's computer.

1. Did May do anything wrong? Why or why not?

2. Is it important to keep hard copy records of data? Why or why not?

Assignment
Due Session 9

Background Information and Hypothesis or Research Question

Identify and summarize the key background information needed to understand your research project. Write these pieces of information as a *bulleted list of statements* (with references!). Your hypothesis or research question should follow from this information and be written at the bottom of the list.

Relevant Background Information

▶

▶

▶

▶

▶

▶

-

-

-

-

Therefore, I hypothesize that

SESSION 9

Defining Your Hypothesis or Research Question

GOALS

Students will

- be able to clearly and concisely define their hypothesis or research question.
- be able to demonstrate understanding of their research project by clearly explaining the background information and experimental results that led to their proposed hypothesis or question.
- reflect on the similarities and differences between research groups.

OUTLINE

Core Activities

Check-in

Presentations of Background and Hypothesis or Research Questions

Announcement—Entering Research Part II

Optional Activities

Discussion of Elements of a Good Hypothesis

Materials for Students

- Projection equipment if students are presenting to entire group
- Index cards

Possible Readings for Session 10

- None

Assignment for Session 10

- Experimental Design and Potential Results with Timeline

FACILITATOR NOTES

Core Activities

Check-in and Peer Lab Visit Reports

Ask students to share one observation about their visit to a peer's research group. They could each highlight one similarity and one difference between their group and their peer's group. Previous students' observations included differences in the following areas:

- research group size
- tidiness of research space
- collaborative vs. individual work
- amount of talking within the group on a daily basis
- repetitive tasks vs. new tasks
- applied vs. basic research
- amount of funding (e.g. make own stock solutions vs. purchasing solutions)

Presentations of Background and Hypothesis or Research Questions

Student presentations can occur either in small groups or in a large group. If in a large group, provide a way for students to project their outline, or make copies of the outline for everyone to follow along during the presentations. Asking students to use the board (chalk talk) is also effective.

Each student should outline the background information leading up to her hypothesis, and then present the hypothesis. A useful way to do this is with a "framing funnel" in which the student presents information in the following format:

1. one word to describe the research (e.g. cancer, nanoparticles)
2. important background information required to understand the research
3. the knowledge gap that the research will address

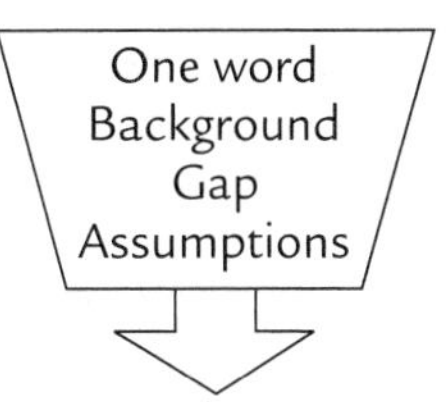

4. assumptions being made (if any) in the research
5. the hypothesis or research question

The framing funnel exercise can be simplified. Students can be asked to describe their research in "one word" and to use that word to start the first sentence of their presentation.

To encourage peer interaction and review, have students write at least one comment, question, or suggestion about each hypothesis on an index card during the presentation. Require each student to verbally present their comment, question, or suggestion for at least two peers during the discussion. At the end of each presentation, collect the index cards and give them to the presenter. Some items for discussion:

- Does the hypothesis "make sense" and follow from the background information?
- Is it clear how testing this hypothesis will contribute to the broader research mission of the student's research group?
- What types of experiments could the student do to test the hypothesis?

Note: During this exercise students frequently realize that their hypothesis or research question is not clearly defined. An effective follow-up exercise is to require them to revise it with their mentor and bring it back the next week.

Announcement

Students are typically making decisions about their next semester's course schedule around this time, so it is a good time to mention Entering Research Part II, which continues to support students as they do research. It focuses on development of a formal presentation (poster or talk) of research results, and, like Part I, should be taken concurrently with independent research credits.

Optional Activities

Elements of a Good Hypothesis

The goal of this discussion is to generate a list of the characteristics of a good hypothesis or research question. Ask students to reflect on the hypothesis or research question presentations just given and brainstorm a list of features of a good hypothesis, either in the large or small groups. At the end, create a comprehensive list of features, which they can use to determine if they should revise their own hypotheses. Some features that students have identified include

- The hypothesis follows logically from the background information.
- The hypothesis is narrowly defined and testable.
- The hypothesis addresses one aspect of a broader research investigation.

Wrap-Up

Assignments for next time:

- Experimental Design and Potential Results with Timeline

Reflections and Notes

STUDENT MATERIALS

To download electronic versions of student materials go to Faculty Lounge (www.whfreeman.com/facultylounge)

Assignment
Due Session 10

Project Design and Potential Results with Timeline

Outline the experiments you will do to test your hypothesis, or the flow and sequence of activities required to complete the project. Include a timeline or flow chart with target dates for the experiments and activities.

For each experiment or activity explain

1. the technique(s) that will be used and the reason(s) for selecting that technique.
2. the type of data that will be collected and why this type of data will inform the hypothesis.
3. all the potential results and whether each would support, or not, your hypothesis. Draw what the predicted results will look like, if applicable (e.g. gel, microscope image, data table).

SESSION 10

Designing Your Experiments

GOALS

Students will

- be able to demonstrate understanding of their research projects by clearly explaining the experimental design.
- be able to demonstrate understanding of their research projects by clearly explaining the techniques they will use and the rationale for using those techniques.
- be able to demonstrate understanding of their research projects by clearly describing the kinds of data they will gather using their experimental techniques and how those data will address their hypotheses or research questions.

OUTLINE

Core Activities

Check-in

Presentation of Experimental/Project Designs (and revised Hypothesis or Research Question)

Research Proposal and Peer Review Assignments

Optional Activities

Discussion of Elements of a Good Experimental Design

Materials for Students

- Computer and projector if students are presenting to entire group
- Index cards

Possible Readings for Session 11

- None

Assignment for Session 11

- Research Proposal Draft #1
- Review of peers' proposal drafts

To Do for Session 11

- Assign students to review two of their peers' proposal drafts.
- Invite mentors, lab members, experienced undergraduates, and students from other sections of Entering Research for final presentations (Sessions 13 and 14).

FACILITATOR NOTES

Core Activities

Check-in

Ask each student to rank how difficult it was to develop his/her experimental design.

Presentation of Experimental/Project Designs (and Revised Hypothesis or Research Question)

Student presentations can occur either in small groups or in a large group. If in a large group, either provide a way for students to project their outlines, or make copies of the outlines for everyone to follow along during the presentations. Each student should restate his/her (revised) hypothesis or research question, and then present the experimental design.

Distribute index cards to the students to write at least one comment, question, or suggestion about the design during the presentation. Require each student to verbally present his or her comment, question, or suggestion to at least two peers during the discussion. At the end of each presentation, collect the index cards and give them to the presenter. Some items for discussion:

- Do the proposed experiments test the stated hypothesis/research question?
- Does the proposed timeline provide sufficient time to complete and analyze data from the proposed experiments?
- Can the student predict potential outcomes of the experiments and explain whether each of those outcomes would support the stated hypothesis?

Research Proposal and Peer Review Assignments

Introduce the research proposal and peer review assignments. Students should have first drafts of their proposals ready for peer review at least 48 hours before the next meeting. The proposal may either be a written document (we recommend 2–3 pages), a scientific poster, or an oral presentation (e.g. PowerPoint).

Optional Activities

Discussion of Elements of a Good Experimental/Project Design

The goal of this discussion is to generate a list of the characteristics of a good research design. Ask students to reflect on the research design presentations just given and brainstorm a list of characteristics of a good research design, either in the large or small groups. At the end, create a comprehensive list of characteristics, which they can use to determine if they should revise their own research designs. Some characteristics that students have outlined in previous offerings include

- The experiments directly address the stated hypothesis.
- Regardless of the outcome, the experimental data will shed light on the validity of the hypothesis.
- The experimental techniques can be completed in the allotted time frame.
- The experimental design is based on the literature.

Wrap-Up

Assignments for next time:

- Research Proposal Draft #1
- Review of peers' proposal drafts

Reflections and Notes

STUDENT MATERIALS

To download electronic versions of student materials go to Faculty Lounge (www.whfreeman.com/facultylounge)

Assignment
Due to your peer reviewer 48 hours before Session 11

Research Proposal Draft #1

Using what you have already written about your project for this workshop series, write a first draft of your research proposal. The following components should be included:

- **Authors:** The student and mentor. Anyone else?

- **Background:** Identify at least three or four key previous findings that lay the foundation for the project and give references from the literature.

- **Hypothesis or Research Question:** Does it follow logically from the background information?

- **Relevance and Implications:** Why is your research important? What may be the potential implications of your results? How will it benefit basic research, human health, or development of a commercial product?

- **Experimental/Project Design:** What experiments will be done to test the hypothesis?

- **Expected Results:** What kind of data would support the hypothesis? What kind would not support it? Have you or others in your research group generated preliminary results that could be included?

- **Acknowledgments:** Who is supporting you in this research (e.g. mentor, other research group members, funding sources)?

Identify at least three aspects of your proposal about which you would like peer review feedback.

1.

2.

3.

Assignment
Draft 1 Review Due Session 11, Draft 2 Review Due Session 12

Peer Review of Research Proposal

Use the rubric below to do an in-depth review of two peers' proposals. In addition, be sure to provide specific feedback on aspects for which the author has requested feedback.

Reviewer ______________________ Author ______________________

	0	1	2	3
Title & Authors	Absent	Title is lengthy and unclear.	Title is lengthy, but clear.	Title is concise and clear.
Background	Absent	The background information presented lacks the content needed to understand the scientific basis of the hypothesis or research question.	The relevant background information is presented, but poorly organized. Therefore, the hypothesis or research question does not follow logically from it.	The relevant background information is presented and is organized such that the hypothesis or research question follows logically from it.
Hypothesis or Research Question	Absent	A statement is made, but it is neither a hypothesis nor a research question.	A hypothesis or research question statement is made, but it is neither concise nor follows logically from the background information.	A clear and concise hypothesis or research question statement is made that follows logically from the background information.
Relevance and Implications	Absent	The stated relevance and implications of the research are general and do not specifically address basic research, human health, or commercial product development.	The stated relevance and implications of the research address only one area of the three mentioned: basic research, human health, or commercial product development.	The stated relevance and implications of the research address two or three of the areas mentioned: basic research, human health, or commercial product development.
Experimental/ Project Design	Absent	Experiments are listed but lack detail and are not connected to the stated hypothesis or research question.	Experiments are listed, and either well explained or connected to the stated hypothesis or research question, but not both.	Experiments are listed, well explained, and connected to the stated hypothesis or research question.
Expected Results	Absent	Potential results are described, but lack a figure to represent them and a statement of whether they would support the stated hypothesis or research question.	Potential results are described, but lack either a figure to represent them or a statement of whether they would support the stated hypothesis or research question.	Potential results are described in a figure and a statement about whether they would support the stated hypothesis or research question is made.
Writing	Absent	Choppy sentence fragments with grammatical and spelling errors.	Mostly clear sentences. Some work needed on clarity and flow. Some grammatical or spelling errors.	Clear. Each sentence deals with one topic and flows logically. No grammatical or spelling errors.

Peer Review Instructions

Please evaluate each component according to the rubric guidelines. Offer ***specific*** suggestions for how to improve the components.

Title & Authors 0 1 2 3
Comments/Suggestions:

Background 0 1 2 3
Comments/Suggestions:

Hypothesis or Research Question 0 1 2 3
Comments/Suggestions:

Relevance and Implications 0 1 2 3
Comments/Suggestions:

Experimental/Project Design 0 1 2 3
Comments/Suggestions:

Expected Results 0 1 2 3
Comments/Suggestions:

Acknowledgments 0 1 2 3
Comments/Suggestions:

Writing (sentence structure, grammar, spelling) 0 1 2 3
Comments/Suggestions:

Comment on the three aspects in the proposal about which the author has requested feedback.

1.

2.

3.

SESSION 11

Research Proposal Review Draft #1

GOALS

Students will:

- understand the importance and role of peer review in the research process.
- appreciate the responsibility of a reviewer.
- share their reviews of their peers' research proposal drafts.

OUTLINE

Core Activities

Check-in

Peer Review Exchange

Optional Activities

None

Materials for Students

- None

Possible Readings for Session 12

- None

Assignment for Session 12

- Discussion Questions:
  - What role does peer review play in the research process?
- Research Proposal Draft #2

To Do for Session 12

- Assign students to review a different set of proposal drafts.
- Invite mentors, lab members, experienced undergraduates, and students from other sections of Entering Research for final presentations (Sessions 13 and 14).

FACILITATOR NOTES

Core Activities

Check-in

Ask each student to identify the part of his/her proposal that was most challenging to write and why.

Peer Review Exchange

Students should get together with reviewers to go over the review comments in general, and specifically the three aspects on which they requested feedback. Encourage them to ask questions and discuss specific strategies for how to improve the draft. Each student should leave with a list of items from the review to discuss with their mentor.

Optional Activities

None

Wrap-Up

Assignments for next time:

- Discussion Questions:
  - What role does peer review play in the research process?
- Research Proposal Draft #2

Reflections and Notes

STUDENT MATERIALS

To download electronic versions of student materials go to Faculty Lounge (www.whfreeman.com/facultylounge)

Assignment
Due ***to your peer reviewer*** *48 hours before Session 12*

Research Proposal Draft #2

Use the comments you received from your first set of peer reviewers, and the suggestions you received from your mentor when you discussed those comments with him/her, to write a second draft of your research proposal.

Again, identify at least 3 aspects of your proposal that you would like your second peer reviewer to give feedback on.

1.

2.

3.

SESSION 12

Research Proposal Review Draft #2

GOALS

Students will

- address the feedback they received on their first proposal draft with their reviewers.

OUTLINE

Core Activities

Check-in

Peer Review

Optional Activities

Discussion of Peer Review

Student Materials

None

Possible Readings for Session 13

- None

Assignment for Session 13

- Final Research Proposal

To Do for Session 13

- Invite mentors, lab members, experienced undergraduates, and students from other sections of Entering Research for final presentations (Sessions 13 and 14).

FACILITATOR NOTES

Core Activities

Check-in

Ask students to share how their discussions with their mentors went about the feedback they received last week.

Peer Review Exchange #2

Students should get together with their second set of reviewers to go over their general comments and the three aspects on which they requested feedback. Encourage them to share the comments of their first reviewers and to highlight the changes they made in response to those comments.

Optional Activities

Discussion of Peer Review

The goal of this discussion is to build understanding and appreciation for the peer review process. In particular, the discussion should address the role that peer review plays in creating shared responsibility for the validity and quality of research generated by the scientific community. Ideas that have emerged in discussion of this question include:

- **What role does peer review play in the research process?**
  - It allows the scientific community to set and enforce standards of research itself.
  - It allows for new interpretations of data.
  - It gives the author a broader perspective on his/her work.
  - It can develop collegiality and a community.
  - It prevents misconduct.

Wrap Up

Assignments for next time:

- ▶ Final Research Proposal

Reflections and Notes

STUDENT MATERIALS

To download electronic versions of student materials go to Faculty Lounge (www.whfreeman.com/facultylounge)

SESSION 13

Final Research Proposal Presentations

GOALS

Students will

- be able to effectively present and defend their research proposals.

OUTLINE

Core Activities

Research Proposal Presentations

Optional Activities

None

Materials for Students

- Research Experience Reflections

Assignment for Session 14

- Research Experience Reflections

FACILITATOR NOTES

Core Activities

Research Proposal Presentations

Each student should formally present his/her research proposal. If students made proposal posters, then this could be a poster session.

Wrap-Up

Assignments for next time:

- Research Experience Reflections

Reflections and Notes

STUDENT MATERIALS

To download electronic versions of student materials go to Faculty Lounge (www.whfreeman.com/facultylounge)

Assignment
Due Session 14

ENTERING RESEARCH PART I

Reflections on Your Research Experience

1. Has your research experience been what you expected it to be so far? Why or why not?

2. What academic and personal goals have you already achieved in your research experience?

3. What values, experiences, and/or perspectives have you contributed to your research team so far? Were you able to contribute in ways that you did not predict? If so, how?

4. What has been the most challenging aspect of your research experience so far and how are you dealing with it?

5. What goals do you have for your research experience in the next semester?

6. What advice would you give to new students interested in having a research experience?

7. How has your research experience prepared you for your future career? Identify the skills you have learned that will be useful in your future career and explain how.

8. Outline at least two differences between research done with a research group and research done as part of a laboratory course.

9. Can you identify at least two connections between your research project and what you are learning in your classes?

10. Based on your experience thus far, how would you modify the letter of recommendation that you drafted in Session 6?

SESSION 14

Final Research Proposal Presentations (continued)

GOALS

Students will

- be able to effectively present and defend their research proposals.

OUTLINE

Core Activities

Research Proposal Presentations

Optional Activities

Post-survey (www.whfreeman.com/facultylounge)

Materials for Students

- Entering Research Part I Student Evaluation
- Post-Survey

FACILITATOR NOTES

Core Activities

Research Proposal Presentations

Each student should formally present his/her research proposal. If students made proposal posters, then this could be a poster session.

Entering Research Part I Student Evaluation

Give students 10–15 minutes to complete the evaluation.

Optional Activities

Post-Survey

If you are participating in the research study, ask students to complete the online survey, the link to which can be found on Freeman and Company's Faculty Lounge web site (www.whfreeman.com/facultylounge).

Wrap-Up

Invite students to share parting thoughts, concerns, or questions.

Encourage students continuing in research to participate in Entering Research Part II.

Reflections and Notes

STUDENT MATERIALS

To download electronic versions of student materials go to Faculty Lounge (www.whfreeman.com/facultylounge)

Entering Research Part I

Evaluation

1. How helpful were these sessions to your learning about and engaging in independent research?

	Do Not Remember	Not Helpful	Somewhat Helpful	Helpful	Very Helpful
Workshop Introduction	○	○	○	○	○
Nature of Science	○	○	○	○	○
Searching for Literature	○	○	○	○	○
Reading Articles & Mentor Styles	○	○	○	○	○
Research Group's Focus	○	○	○	○	○
Expectations & Goals	○	○	○	○	○
Who's Who in Your Group	○	○	○	○	○
Documenting Your Research	○	○	○	○	○
Defining Hypothesis	○	○	○	○	○
Designing Experiments	○	○	○	○	○
Peer Review Draft #1	○	○	○	○	○
Peer Review Draft #2	○	○	○	○	○
Final Proposal Presentation	○	○	○	○	○

Please comment on any of the above sessions.

2. How helpful were these assignments in supporting your independent research experience?

	Do Not Remember	Not Helpful	Somewhat Helpful	Helpful	Very Helpful
Finding a Research Mentor	○	○	○	○	○
Research Exp. Expectations	○	○	○	○	○
Your Research Group's Focus	○	○	○	○	○
Mentor Biography	○	○	○	○	○
Expectations Disc. Summary	○	○	○	○	○
Scientific Article Worksheet	○	○	○	○	○
Research Group Diagram	○	○	○	○	○
Visiting Peer Research Groups	○	○	○	○	○
Background Info & Hypothesis	○	○	○	○	○
Exp. Design & Potential Results	○	○	○	○	○
Research Proposal Draft #1	○	○	○	○	○
Research Proposal Draft #2	○	○	○	○	○
Peer Review	○	○	○	○	○
Research Exp. Reflections	○	○	○	○	○
Proposal Presentation	○	○	○	○	○

Please comment on any of the above assignments.

3. How effective were each of these class formats used in the workshop series?

	Do Not Remember	Not Effective	Somewhat Effective	Effective	Very Effective
In-class Discussions	○	○	○	○	○
On-line Discussions	○	○	○	○	○
Case Studies	○	○	○	○	○
Role Play Exercises	○	○	○	○	○
In-class Activities	○	○	○	○	○
Student Disc. Facilitation	○	○	○	○	○
Instructor Disc. Facilitation	○	○	○	○	○

Please comment on any of the above class formats.

4. Any additional comments about the workshop series?

PART II

ENTERING RESEARCH
Session by Session

Facilitator Notes and Materials for Students

SESSION 15

Introduction to the Workshop Series and Science Communication

GOALS

Students will

- become familiar with Entering Research Part II content, structure, and learning objectives.
- begin to form a learning community and establish ground rules for discussions.
- share their motivation for doing research and their research project with their peers.
- discuss science as a way of thinking and how science is communicated to different audiences.

OUTLINE

Core Activities

Introductions

Entering Research Part II Overview

- Syllabus
- Grading
- Assignments

Discussion of Research Experience Expectations

Optional Activities

Pre-survey (www.whfreeman.com/facultylounge) Discussion of Science Communication

Materials for Students

- Name plates
- Constructive/Destructive Group Behaviors Handout (appendix)
- Syllabus, including learning objectives
- Student Discussion Guidelines (Session 1)
- Abstracts Handout
- Research Project Outline and Example Scientific Abstract Assignment
- Researching Research Careers Assignment

Possible Readings for Session 16

- Slaughter, G. R. (2006) Chapter 9—**Writing Scientific Abstracts**, *Beyond the Beakers: Smart Advice for Entering Graduate Programs in the Sciences and Engineering*, Baylor College of Medicine. www.bcm.edu/gs/BeyondTheBeakers/Table%20of%20Contents.htm

Assignments for Session 16

- Research Project Outline and Science Abstract

FACILITATOR NOTES

If Entering Research Part II is being offered as a continuation of Entering Research Part I, and the students are familiar with the workshop content and format, the facilitator may modify or eliminate many of the activities outlined below. If some students are new and others are continuing, the facilitator may ask the returning students to introduce the course to the new students.

Core Activities

Introductions

Building a strong, safe learning community should begin on the first day of class. See the Session 1 facilitator notes for specific icebreaker activity ideas and mechanisms for establishing confidentiality and discussion ground rules.

Entering Research Part II Workshop Series Overview

This workshop series is designed to compliment and provide support for students continuing independent research in the sciences. **The main goal of Entering Research Part II is to help students prepare and present a formal presentation of their research results.**

Syllabus and Learning Objectives

Hand out and briefly review the Entering Research Part II syllabus and learning objectives. A sample syllabus can be found in the appendix.

Student Information Sheet

Hand out and ask students to fill in information about their research mentors.

Grading

Discuss the grading scheme. Below is a guide that aligns with the leaning objectives presented in this manual. However, each instructor should design a grading scheme that aligns with the learning objectives outlined in his/her own syllabus

10% Attendance

10% In-Class Participation

10% Project Outline and Scientific Abstract

5% General Public Abstract

20% Peer Reviews (10% each for Abstract and Poster)

25% Presentation (Poster or Talk)

20% Mini-grant Proposal

Assignments

Give an overview of the class assignments, all of which help students develop presentations of their research. The assignments build on one another and increase in complexity. Starting with a few sentences in the Outline, to an Abstract, a Poster, and finally a Mini-grant Proposal in which they propose the next step in their research. Reassure them that the components will be developed with support and revisions, both from class peers and outside reviewers (scientists/graduate students). If there is a public venue (e.g. University-wide Undergraduate Research Symposium) at which students can present their research presentations, in addition to their in-class presentation, then either encourage or require them to participate in it.

Research Experience Expectations Discussion

The goal of this discussion is to help students get to know one another and their motivations for doing research. Break students into small groups to discuss the questions, then bring the large group back together to share one thing from their discussion, such as something they all had in common or something they learned from their peers. Point out how the workshops will offer support to address concerns that are raised.

Discussion questions:

- Why are you doing research?
- What do you hope to gain from the Entering Research workshops?
- What has been your greatest challenge and greatest success in research so far?
- What specific goals do you hope to achieve this semester on your research project?

Optional Activities

Pre-survey

If you are participating in the research study, ask students to complete the online survey, the link to which can be found on Freeman and Company's Faculty Lounge web site (www.whfreeman.com/facultylounge).

Science Communication Discussion

If the workshop series is a continuation of Entering Research Part I, then there will likely be time to have a general discussion about science communication. Questions to guide this discussion include

- How do scientists communicate their research to one another? To the public?
- Whose job is it to communicate science to the public?
- What role do the popular media play in communicating science?
- How do scientists convey research accurately, yet understandably?

Science Careers

Ask students to share the careers they are interested in pursuing, and introduce the Researching Research Careers Assignment. Each student should identify a career she would like to learn more about through the research careers interview assignment. If more than one student will be exploring the same career, they should coordinate their efforts so as not to interview the same person twice.

Wrap-Up

Assignments for next time:

- Research Project Outline and Science Abstract
- Researching Research Careers (Due session 22)

Reflections and Notes

STUDENT MATERIALS

To download electronic versions of student materials go to Faculty Lounge (www.whfreeman.com/facultylounge)

Handout

Entering Research Part II: Student Information Sheet

	Student	**Mentor #1 Name & Email** (Professor or principal investigator)	**Mentor #2 Name & Email** (e.g. grad student, post-doc)
1.			
2.			
3.			
4.			
5.			
6.			
7.			
8.			
9.			
10.			
11.			
12.			

Handout

Abstracts

Modified from "*A Guide to Writing in the Sciences*" by Gilpin and Patchet-Golubev, 2000.

An abstract is a brief yet comprehensive summary of a research report, without added interpretation or criticism. It conveys the most significant information about the research, especially the results. Readers often decide, on the basis of the abstract, whether or not to read the full article.

The abstract should

1. be concise, but not highly abbreviated;
2. assume the reader has some knowledge in the subject area;
3. state the research question or hypothesis;
4. present the approach used to answer the question;
5. report the most important results (but not cite figures or tables); and
6. indicate the main conclusion(s) near the end.

Some key phrases to use when writing an abstract include

The main purpose of the study was to . . .

To address this question we used . . . approach (rationale)

The main findings suggest that . . .

Therefore, these results support the conclusion that . . .

Assignment
Due Session 16

Research Project Outline and Science Abstract

Research Group's Focus

Research Project Title

Introduction/Background

Identify and summarize the key background information needed to understand your research project. Write these pieces of information as a bulleted list of statements. Your hypothesis or research question should follow from this information.

▶

▶

▶

▶

▶

▶

▶

▶

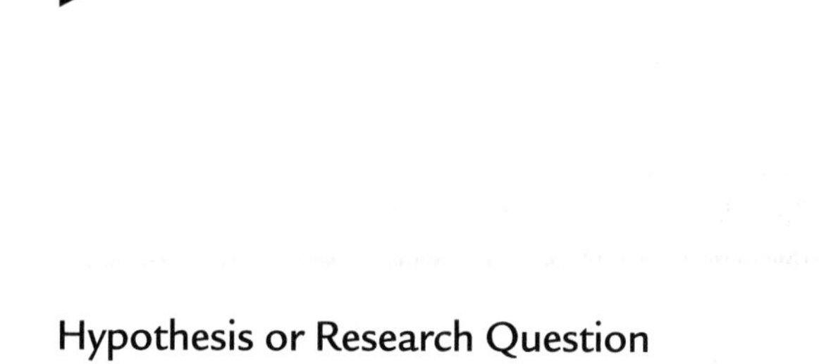

Hypothesis or Research Question

Relevance and Implications of Your Research Project

Why is your research important? What may be the potential implications of your results? How will it benefit basic research, human health, or development of a commercial product?

Experimental Design and Potential Results

Outline the experiments you will do to test your hypothesis. For each experiment explain

1. the technique(s) that will be used and the reason(s) for selecting that technique.

2. the type of data that will be collected and why this type of data will inform the hypothesis.

3. (optional) all the potential results and whether each would support, or negate, your hypothesis. Draw what the predicted results will look like, if applicable (e.g. gel, microscope image, data table or graph).

Timeline

Outline a weekly or monthly timeline for your project. Be sure to refer to each of the proposed experiments (or parts of the experiments), allow time for analysis of data, and allow time for the preparation of a presentation of the data (e.g. poster or oral presentation).

Abstract

Synthesize the core information in your outline and write a prospective scientific abstract of 200 words or less.

Handout

Example of Research Project Outline and Science Abstract

by Jennifer Arens-Gubbels, University of Wisconsin—Madison

Research Group's Focus

Epithelial Ovarian Cancer: Metastasis and Immune Evasion

Research Project Title

The role of cell surface MUC16 contributing to immune evasion in epithelial ovarian cancer

Introduction/Background

- Ovarian cancer is the most deadly of gynecological cancers
- Cells from the tumor of this cancer express large amounts of a protein called MUC16. This protein is in the mucin family and is both secreted into the peritoneal fluid and expressed on the surface of tumor cells. MUC16 is a very large, heavily glycosylated protein, with a molecular weight of over 3 million daltons. It is expressed normally sparsely on the surface of the ovary as well as the endometrium.
- Because this molecule is so large in structure and heavily glycosylated, it may protect the tumor cell from immune attack via steric hinderance.
- Natural killer (NK) cells kill tumors and viruses, and require close proximity to their targets in order to effectively kill a tumor cell.

Hypothesis or Research Question

We hypothesize that the large molecular weight of MUC16 provides a "cloak" to protect ovarian cancer tumor cells from attack by NK cells.

Relevance and Implications of Your Research Project

- **Basic science:** Establishing the physiological relationship between MUC16 and NK cells in immune evasion of epithelial ovarian cancer cells will provide a forum for investigating the role of mucins in signal transduction pathways that modulate cancer cell behavior.
- **Human health:** Understanding how MUC16 mediates immune cell evasion has implications for the prevention, early detection, and treatment of epithelial ovarian cancer.
- **Commercial product:** not applicable for this project

Experimental Design and Potential Results

- In order to determine if NK cells can form activating immune synapses (defined as close contact with a target cell and polarization of activating receptors at that interface) with tumor cells, we will incubate NK cells and OVCAR-3 (ovarian cancer cell line with MUC16) or #7 cells (OVCAR-3 derived cell line that does not produce MUC16) together for 25 minutes.
- We will then put the cells on a coverslip, fix them, and stain for the following markers: F-actin, LFA-1, CD2, or perforin. F-actin will stain all cell types, but LFA-1, CD2, and perforin will be present only on the NK cells.
- We will count the number of functional synapses that the NK cells are able to form with either cell type using fluorescent microscopy. We will define an activating immune synapse by the polarization of any of the four markers at the contact point between the NK cell and the tumor cell.
- If our hypothesis is true, then the MUC16– cells will form more synapses with NK cells compared to the MUC16+ cells in this experiment.

Timeline

- Weeks 1 and 2: order supplies and standardize coverslip procedure
- Weeks 3, 4, 5 and 6: conduct synapse experiments with four healthy donors
- Weeks 7, 8, 9 and 10: count synapses and graph/analyze data
- Week 11, 12, 13 and 14: repeat synapse experiments (if other experiments did not work properly, or if more data points are needed for statistical analysis)
- Week 15 and 16: analyze final data and write up results or prepare presentation

Scientific Abstract

Epithelial ovarian cancer is the deadliest of gynecological cancers. Cells of this tumor type overexpress MUC16, a heavily glycosylated protein with a molecular weight of over 3 million Da. This protein is expressed on the surface of ovarian tumor (OT) cells as well as secreted into the peritoneal fluid of ovarian cancer patients. We have shown previously that soluble MUC16 can bind to natural killer (NK) cells and cause them to become non-cytotoxic. Therefore, we decided to investigate the effect of cell surface MUC16 on the ability of NK cells to lyse OT cells. To implement these experiments, we utilized an OT cell line that expresses MUC16 (#12 cells) and a knock-down line that lacks MUC16 (#7 cells). The ability of NK cells to form activating immune synapses (AIS) with either cell type was investigated using confocal microscopy. NK cells and either #7 or #12 cells were incubated together and stained with dyes to visualize several AIS proteins. We determined that NK cells form more AIS when incubated with #7s, compared to #12s. Our results indicate that cell surface MUC16, because of its large size and heavily glycosylated regions, presents a steric barrier to NK cells and prevents AIS formation.

Assignment
Due Session 22

Researching Research Careers

Many of the skills learned doing research are important when preparing for a research career, and also for many other careers. Below is a list of careers for which research training is important. Each member of the class should select a different career on the list (or one that is not on the list), identify an individual with this career, and do an email interview of that person. An email template is given below. **Set up an interview at least 1 week before this assignment is due and email the name to your facilitator.**

Careers

- Museum professional
- Entrepreneur
- Editor
- Science writer
- Patent lawyer
- Government scientist
- Science policy advisor
- Private industry scientist
- Research university professor
- Research university professional staff (researcher, instructor)
- Teaching university/college professor
- Outreach coordinator (private or academic)
- Clinical researcher (e.g. clinical chemist, hospital clinic manager)
- Academic or private administrator/leader
- Others ___________________

Email Template

Dear Dr./Mr./Ms.____________,

I am a student at __________________ and am writing to request an email interview for my Entering Research class. We are studying different careers that research training can prepare us for, and I am interested in your career as a _______________. If you are willing to answer a few questions, I would really appreciate it. If you do not have the time, perhaps you could forward this to a colleague who might.

Sincerely, _________________

Questions:

1. What do you do in your job?
2. What kind of education or training is needed for your career? Is research training needed? Why or why not?
3. What is a typical starting salary in your career?
4. How much time do you have for personal, non-career interests?
5. What advice do you have for young people interested in pursuing your career as __________________?

SESSION 16

Research Project Outlines and Scientific Abstracts

GOALS

Students will

- use their project outline to describe their research project to the class.
- understand their peers' research projects.

OUTLINE

Core Activities

Check-in

Presentation of Research Project Outlines and Scientific Abstracts

NOTE: Depending on how many students are in the class, it may take two sessions to get through all of the project outline presentations. Session 17 has also been dedicated to this, but may focus solely on mentor–mentee relationships if not needed to finish project outline presentations.

Optional Activities

Revisit Discussion of Elements of a Good Hypothesis (Session 9)

Materials for Students

- Research Project Outline Peer Feedback Form
- Reflecting on Your Mentoring Relationship

Possible Readings for Session 17

- None

Assignment for Session 17

- Finish research project outline and scientific abstract presentations
- Reflecting on Your Mentoring Relationship

FACILITATOR NOTES

Core Activities

Check-in

Ask each student to share his/her greatest success and biggest challenge as a researcher (so far).

Research Project Outline Presentations

Ask students to use their outlines to explain their research project to their peers in chalk talks. Allow a few minutes for the students to review their peers' outlines and **5–10 minutes per student** (5–6 students total) for the presentation. A good way to help students structure their talks is to ask them to begin by completing this sentence

"I am interested in understanding/determining/developing. . . . "

and to end by saying

"If everything goes well, my research will. . . . "

During the presentations, the other students should complete a peer review form, including at least one question or comment to share with the presenter. Students should submit their forms to the presenter at the end of class.

This activity provides a great opportunity for students to explain terminology (jargon) that is specific to their area of research. In the process, they build confidence when they realize they are able to accurately use this specific terminology.

Optional Activities

Discussion of Elements of a Good Hypothesis (Revisit Session 9)

The goal of this discussion is to remind students of the elements of a good hypothesis and to offer them the opportunity to reflect on, and possibly reconsider, their own hypothesis. This discussion can happen after the presentations for the day are complete, or be interwoven between presentations.

Discussion questions:

- What is a research hypothesis/question?
- How do you evaluate one? What makes it good or bad?
- What is the difference between a hypothesis/research question and a research project?

Wrap-Up

Assignments for next time:

- Continue research project outline presentations
- Reflecting on Your Mentoring Relationship

Reflections and Notes

STUDENT MATERIALS

To download electronic versions of student materials go to Faculty Lounge (www.whfreeman.com/facultylounge)

Handout

Research Project Outline Peer Feedback Form

Presenter ______________________

1. Based on the project outline, scientific abstract, and oral presentation, how well do you understand this research project?

   ○ no understanding at all

   ○ weak understanding

   ○ good understanding

   ○ very good understanding

2. In your own words, write one or two sentences to describe the main focus of this research project.

3. Write at least one question or comment for the presenter

Assignment
Due Session 17

Reflecting on Your Mentoring Relationship

Maintaining a positive relationship with your research mentor is very important and can be achieved through frequent, open, and candid communication. To facilitate this communication, answer the questions below, then meet with your mentor to discuss them. You may also give a copy of the questions to your mentor to reflect on before the meeting.

1. What seems to be working well for you in the mentor–mentee relationship?

2. What is not working so well for you?

3. Review the goals and expectations you established with your mentor at the beginning of your relationship. Do you still agree that these goals and expectations are appropriate for your research experience, or do they need to be adjusted? Are you satisfied with rate of progress you have made toward reaching the goals? If not, what might you do differently?

4. What has the relationship you have with your mentor taught you about what you must do to be successful as a researcher?

5. What aspects of mentoring do you need to get from someone other than your direct mentor? Who can provide this mentoring?

Write a paragraph summarizing the conversation you had with your mentor.

SESSION 17

Research Project Outlines and Scientific Abstracts (continued)

GOALS

Students will

- use their project outline to describe their research project to the class.
- understand their peers' research projects.

OUTLINE

Core Activities

Check-in

Continue from Session 16: Presentation of Research Project Outlines and Scientific Abstracts

Optional Activities

Discussion of Reflecting on Your Mentoring Relationship

Prioritizing Research Mentor Roles

Materials for Students

- Research Project Outline Peer Feedback Form (Session 16)

Possible Readings for Session 18

- Derry, G. N. (1999) Chapter 10—**More Practical Questions: Science and Society**, *What Science Is and How It Works*, Princeton University Press
- Gregory, J. and Miller, S. (1998) Chapter 1—**The Recent "Public Understanding of Science Movement,"** *Science in Public: Communication, Culture, and Credibility*, Basic Books.

Assignment for Session 18

- Discussion Questions:
  - What should people know about science?
  - Is it important for the general public to understand your research?
  - How will you make your research presentation accessible to the general public?

FACILITATOR NOTES

Core Activities

Check-in

Ask each student to rate how difficult it was to have the "relationship" discussion with his/her mentor.

Research Project Outline Presentations (cont.)

Ask students to use their outlines to explain their research project to their peers. Allow **5–10 minutes per student** (5–6 students total). During the presentations, the other students should complete a peer review form, including at least one question or comment to share with the presenter. Students should submit their forms to the presenter at the end of class.

Optional Activities

Discussion of Reflecting on Your Mentoring Relationship

The goal of this discussion is to give students the opportunity to share the conversations they had with their mentors about their mentor–mentee relationship.

Discussion questions:

- Was it difficult to have this conversation with your mentor? Why or why not?
- As a result of this conversation, have you and your mentor decided to change the way you are working together? If so, how?
- Ideally, how frequently would you like to have conversations like this with your mentor? Why?

Prioritizing Research Mentor Roles

The goal of this activity is to help students identify the different roles that research mentors can play and to prioritize those roles based on their needs.

Ask students to

- Individually rank the roles they hope their research mentor will play.
- Pair up and explain their top two or three.
- Create a large group summary of the top two roles.
- Do all of these roles need to be fulfilled by ONE (research) mentor? Who else in your research group, or beyond, could be a mentor for you?

Wrap-Up

Assignments for next time:

- Discussion Questions
  - What should people know about science?
  - Is it important for the general public to understand your research?
  - How will you make your research presentation accessible to the general public?

Reflections and Notes

STUDENT MATERIALS

To download electronic versions of student materials go to Faculty Lounge (www.whfreeman.com/facultylounge)

Activity

Roles for Your Research Mentor

Consider the different roles of research mentors listed below. Add additional roles that may be missing from the list. Prioritize these roles according to your expectations, with #1 as the most important.

Role	Priority
Teach by example	
Train you in disciplinary research	
Improve your writing and communication skills	
Provide growth experiences	
Help build your self-confidence as a researcher	
Model and promote professional behavior	
Inspire	
Offer encouragement	
Assist with advancement of your career	
Facilitate networking with colleagues	
Help build the bridge between research and clinical work	
Other:	
Other:	
Other:	

SESSION 18

Science and Society

GOALS

Students will

- become aware of how science and society interact.
- consider the societal implications of their research.
- recognize their responsibility to communicate their research to the general public.
- develop strategies to "translate" their research to the general public.

OUTLINE

Core Activities

Check-in

Discussion of Science and Society

Optional Activities

Science Literacy Test

Materials for Students

- Science Literacy Test

Possible Readings for Session 19

- Handrix, M.J.C. (April 2001) **Communicating Science**, *FASEB News*

Assignment for Session 19

- General Public Abstract
- Review of two peers' abstracts

FACILITATOR NOTES

Core Activities

Check-in

Ask students to identify one way their research project is relevant to society.

Discussion of Science and Society

The goal of this discussion is to encourage students to consider the research they are doing from the perspective of the general public. They should incorporate what they learned from their reading into the discussion. Ideas that have emerged in discussion of these questions in previous offerings include:

- **What should people know about science?**
  - Basic science concepts
  - The process of research
- **Is it important for the general public to understand your research? Why or why not?**
  - My research is very basic, so they don't really need to understand it.
  - My research is applied and could potentially impact people's daily lives, so they should understand it.
- **How will you make your research presentation accessible to the general public?**
  - I will avoid using scientific jargon.
  - I will use images and drawings whenever I can to explain my research.
  - I will focus on the "BIG" research question and not the specific details of my experiments.

Optional Activities

Science Literacy Test

Give students about 10 minutes to take the written part of the science literacy test, another 5 minutes to compare their answers with a classmate, and then go over the answers.

The interview questions in the second part of the science literacy test can be used for discussion, in addition to the following questions.

- What is scientific literacy?
- Does the public need to be scientifically literate?
- Does this test accurately assess scientific literacy?
- Is it more important for the general public to know scientific facts/content, or for them to understand the scientific process? Why?
- What questions would you ask on a scientific literacy test?

View a Short Film

Students can be introduced to the value of science literacy by viewing and discussing a short film, such as "A Private Universe," that was made to highlight science literacy issues.

Wrap-Up

Assignment for next time:

- General Public Abstract
- Review of two Peers' Abstracts

Reflections and Notes

STUDENT MATERIALS

To download electronic versions of student materials go to Faculty Lounge (www.whfreeman.com/facultylounge)

Activity

Scientific Literacy Test

Written Questions

T/F The center of the Earth is very hot.

T/F All radioactivity is man-made.

T/F It is the father's gene that decides whether the baby is a boy or a girl.

T/F Lasers work by focusing sound waves.

T/F Electrons are smaller than atoms.

T/F Antibiotics kill viruses as well as bacteria.

T/F The universe began with a huge explosion.

T/F The continents have been moving their locations for millions of years and will continue to move.

T/F Human beings have developed from earlier species of animals.

T/F The sun goes around the earth.

Note: The answer key and current data regarding the performance of different groups in answering these questions are available on the National Science Foundation's Science and Engineering Indicators web site: www.nsf.gov/statistics/seind06/

Interview Questions

1. When you read news stories, you see certain sets of words and terms. We are interested in how many people recognize certain kinds of terms, and I would like to ask you a few brief questions in that regard. First, some articles refer to the results of a scientific study. When you read or hear the term scientific study, do you have a clear understanding of what it means, a general sense of what it means, or little understanding of what it means? If the response is "clear understanding" or "general

sense," in your own words, could you tell me what it means to study something scientifically?

2. Now, please think of this situation: Two scientists want to know if a certain drug is effective in treating high blood pressure. The first scientist wants to give the drug to 1,000 people with high blood pressure and see how many experience lower blood pressure levels. The second scientist wants to give the drug to 500 people with high blood pressure and not give the drug to another 500 people with high blood pressure and see how many in both groups experience lower blood pressure levels. Which is the better way to test this drug? Why is it better to test the drug this way?

3. Now think about this situation: A doctor tells a couple that their "genetic makeup" means that they've got one in four chances of having a child with an inherited illness. Does this mean that if their first child has the illness, the next three will not? Does this mean that each of the couple's children will have the same risk of suffering from the illness?

Assignment
Due to peer reviewers 24 hours before Session 19.

Draft of General Public Abstract

1. Modify your scientific abstract for the general public (200 words or less). To help you think about the modifications you need to make, imagine you are sitting on an airplane and the person sitting next to you asks, "What do you do?" How would you explain your research to this person?

2. When presenting research to the general public you may be asked about ethical issues associated with your work. Identify at least two ethical issues associated with your research and explain how you would address those issues to a general audience.

Handout

Example of General Public Abstract

by Jennifer Arens-Gubbels, University of Wisconsin–Madison

General Public Abstract

Epithelial ovarian cancer is the deadliest of gynecological cancers. Cells of this tumor type express an extremely large molecule called MUC16. This molecule is found covering the surface of the tumor cells, as well as floating in elevated amounts in the body fluids of ovarian cancer patients. We have shown previously that MUC16 in body fluid binds to immune cells and causes them to be non-functional. We therefore investigated the effects of cell surface MUC16 to determine if it also functions to inhibit immune cells. We incubated ovarian cancer cells that express MUC16 or ovarian cancer cells that lack MUC16 with immune cells to determine if the immune cells could create an activating immune synapse with the cancer cells. Activating immune synapses are indicators of target cell death, require cell-to-cell contact, and give detailed information about the functionality of immune cells. We found that cells that lack MUC16 form more activating synapses with immune cells compared to cells that have large amounts of MUC16 on their surface. Our results indicate that MUC16, because of its large size, may be physically inhibiting immune cells from forming functional activating immune synapses.

Ethical Issues

- **Research Ethics:** There could be bias in counting synapses if the cell type that the NK cells were incubated with is known. To address this we used an experimental design in which the person counting the synapses was blind to the cell type. In addition, another member of the lab, who is unfamiliar with the project, also counted the synapses to ensure accurate counts were taken.
- **Bioethics:** Cells used in this research are derived from human donors. Appropriate legal procedures were followed in harvesting and using the cells, which include full consent from the donor of the cells.

Assignment
Due Session 19

General Public Abstract Peer Review Form

Complete reviews for two peers' draft abstracts. Bring the reviews to class prepared to discuss them with the authors in small groups.

Reviewer ______________________ Author ______________________

Please evaluate each section and offer ***specific*** suggestions for how to improve the components.

Title and Authors
Comments/Suggestions

Hypothesis or Research Question
Comments/Suggestions

Context or Relevance
Comments/Suggestions

Research Methods
Comments/Suggestions

Major Results
Comments/Suggestions

Conclusions, including Relevance
Comments/Suggestions

Writing (sentence structure, grammar, spelling)
Comments/Suggestions

SESSION 19

Peer Review of General Public Abstracts

GOALS

Students will

- be able to identify the attributes of good public abstracts.
- be able to give constructive feedback to their peers on their general public abstracts.

OUTLINE

Core Activities

Check-in

General Public Abstract Peer Review

Optional Activities

Discussion of common problems/solutions and best features of public abstracts

Materials for Students

Ethics Case with Mentor Assignment

Possible Readings for Session 20

- Couzin, J. (September 2006) "**Truth and Consequences**," *Science*, 313: 1222–1226.
- Slaughter G. R. (2006) Chapter 7—**Avoiding Plagiarism/Copyright Infringement**, *Beyond the Beakers: Smart Advice for Entering Graduate Programs in the Sciences and Engineering, Baylor College of Medicine.* http://www.bcm.edu/gs/BeyondTheBeakers/Table%20of%20Contents.htm
- Derry, G. N. (1999) Chapter 11—**Difficult and Important Questions: Science, Values, and Ethics**, *What Science Is and How It Works*, Princeton University Press.

Assignment for Session 20

- Final draft of General Public Abstract
- Ethics Discussion with Mentor

- Discussion Questions:
  - Do you agree with your mentor's response to the scientific misconduct case you shared with them? Why or why not?
  - Why is misconduct such an important issue in the scientific community?
  - What measures are in place to help prevent misconduct?

FACILITATOR NOTES

Core Activities

Check-in

Ask students to rate how difficult it was to modify their scientific abstract for the general public.

General Public Abstract Peer Review Exchange

Students will gather with their peers to give one another feedback on their abstract drafts. In addition to sharing review forms, reviewers should specifically

1. identify the part of the abstract where the importance of the research to the general public is presented, and
2. identify words in the abstract that are "scientific," and likely not part of the vocabulary of a non-scientist.

The focus of the discussion should be on how to clarify or better convey the importance of the research, and how to replace the "scientific" terminology with words or phrases better suited for a general audience.

Ask students to generate a list of common problems encountered in their drafts, with a list of the strategies they came up with to address those problems. In addition, they can generate a list of the good features of their abstracts, which will lead into the optional discussion activity outlined below.

Optional Activities

Discussion of Common Problems (and Solutions) and Best Features of Public Abstracts

The goal of this discussion is to compile the ideas generated in small groups about how to improve their abstracts. Each group can share its lists of problems/solutions and good features with the large group. The facilitator should ask the groups to take turns sharing one thing from their list until all items have been presented. Writing these on the board as they are presented will create a compiled list for the students to keep for future reference.

Wrap-Up

Assignments for next time:

- Final draft of General Public Abstract
- Ethics Case with Mentor
- Discussion questions:
  - Do you agree with your mentor's response to the scientific misconduct case? Why or why not?
  - Why is misconduct such an important issue in the scientific community?
  - What measures are in place to help prevent misconduct?

Reflections and Notes

STUDENT MATERIALS

To download electronic versions of student materials go to Faculty Lounge (www.whfreeman.com/facultylounge)

Assignment
Due Session 20

Final Draft of General Public Abstract

Revise your general public abstract based on the feedback received from peer reviewers.

Assignment
Due Session 20

Ethics Case Discussion with Mentor

Discuss this case study with your mentor and ask him/her to remember what it was like to be an undergraduate student. Write a one-paragraph summary of the conversation you had about it.

Too Good to Be True?

Evelyn and John joined the lab at the same time as sophomores and have been doing research on related, yet separate projects for the past year. Evelyn, a quiet and very diligent worker, spends many hours in the lab working on her project. She has encountered several obstacles in her research, but is making slow, yet consistent progress. She sees John there infrequently and notices that he spends most of his time chatting with the other lab members. The PI of the lab travels a lot, but when he is there, John always seems to connect with him.

At lab meeting last week, John presented his research. The results he reported were exactly what the PI was looking for. The PI was ecstatic. Evelyn was stunned. She does not remember seeing John do any of the experiments he presented. She suspects that he is not being truthful, but has no proof. His research is linked to hers, so if the results are not valid, it will negatively impact her project, and the entire lab. Everyone really likes John, including the PI, and everyone knows that she has been dealing with a lot of set backs in her research. She doesn't want to look like a jealous co-worker by accusing John of fabricating data, but she truly suspects that he has. What should she do?

SESSION 20

Research Ethics

GOALS

Students will

- be aware of and able to explain the importance of ethical conduct in research.
- be able to engage in a discussion about misconduct in research.

OUTLINE

Core Activities

Check-in

Discussion of Ethics Case with Mentor

Ethics case study and/or role-play exercise

Optional Activities

Discussion of Truth or Consequences Article

Materials for Students

- Case studies and/or role play activities
- Scientific Poster Hunt Assignment

Possible Readings for Session 21

- Slaughter, G. R. (2006) Chapter 10—**Creating Effective Figures**, *Beyond the Beakers: Smart Advice for Entering Graduate Programs in the Sciences and Engineering*, Baylor College of Medicine. www.bcm.edu/gs/BeyondTheBeakers/Table%20of%20Contents.htm
- Slaughter, G. R. (2006) Chapter 11—**Posters with Pizzazz**, *Beyond the Beakers: Smart Advice for Entering Graduate Programs in the Sciences and Engineering*, Baylor College of Medicine. www.bcm.edu/gs/BeyondTheBeakers/Table%20of%20Contents.htm
- Slaughter, G. R. (2006) Chapter 12—**Oral Presentations**, *Beyond the Beakers: Smart Advice for Entering Graduate Programs in the Sciences and Engineering*, Baylor College of Medicine. www.bcm.edu/gs/BeyondTheBeakers/Table%20of%20Contents.htm

Assignment for Session 21

- Scientific Poster Hunt

To Do for Session 22

- Invite an experienced researcher to present a research presentation (poster or oral) for Session 22.

FACILITATOR NOTES

Core Activities

Check-in

Ask students why they think ethics is important in research.

Discussion of Ethics Case with Mentor

The goal of this discussion is to share mentors' responses to the "Too Good to Be True?" case study. Ask students to reflect on the structure of their research group and consider to whom they would go if they were in this situation. How important are the social dynamics in a research group? Ideas that have emerged in discussion of these questions in previous offerings include

- **Do you agree with your mentor's response to the scientific misconduct case? Why or why not?**
  - The reality of research group interactions is more complex than I thought.
  - My mentor said Evelyn should just come to him and everything would work out, but I don't know if I believe this.
- **Why is misconduct such an important issue in the scientific community?**
  - Dishonesty can lead to corruption.
  - Small lies or fabrications can have large impacts because future work is based on them.
  - It can waste taxpayer money.
  - It defeats the whole purpose of doing research.
- **What measures are in place to help prevent misconduct?**
  - Peer review
  - Replicating experiments
  - Notebooks/documenting steps

Ethics Case Studies and Role Play

Three scenarios modified from the National Academy's "On Being a Scientist: Responsible Conduct in Research" publication (1995) are presented in the student materials. Any may be used to generate discussion.

- Case: The Selection of Data
- Role Play: The Sharing of Research Materials
- Case: Credit Where Credit is Due

Optional Activities

Discussion of "Truth or Consequences" Article

The goal of this discussion is to explore a "real life" incident of academic misconduct and the impact it had on the people involved. Students should read this article before class (Session 19).

Discussion questions:

- What would you do if you suspected your PI or mentor of falsifying data?
- Would the stage of your academic career (e.g. sophomore undergraduate vs. fourth-year graduate student) impact your decision about what you would do? How?

Wrap-Up

Assignments for next time:

- Scientific Poster Hunt

Reflections and Notes

STUDENT MATERIALS

To download electronic versions of student materials go to Faculty Lounge (www.whfreeman.com/facultylounge)

Activity

Ethics Case: The Selection of Data

(Modified from "On Being a Scientist: Responsible Conduct in Research," 2nd ed., National Academy Press, 1995)

Seniors Deborah and Kathleen have made a series of measurements on a new experimental semiconductor material using an expensive neutron source at a national laboratory. When they get back to their own lab and examine the data, they get the following data points. A newly proposed theory predicts results indicated by the curve.

During the measurements at the national laboratory, Deborah and Kathleen observed that there were power fluctuations they could not control or predict.

Furthermore, they discussed their work with another group doing similar experiments, and they knew that the group had gotten results confirming the theoretical prediction and was writing a manuscript describing their results.

In writing up their own results for their senior research project and hopefully for publication, Kathleen suggests dropping the two anomalous data points near the abscissa (the solid squares) from the published graph and from the statistical analysis. She purposes that the existence of the data points be mentioned in the paper as possibly due to power fluctuations and being outside the expected standard deviation calculated from the remaining data points. "These two runs," she argues to Deborah, "were obviously wrong."

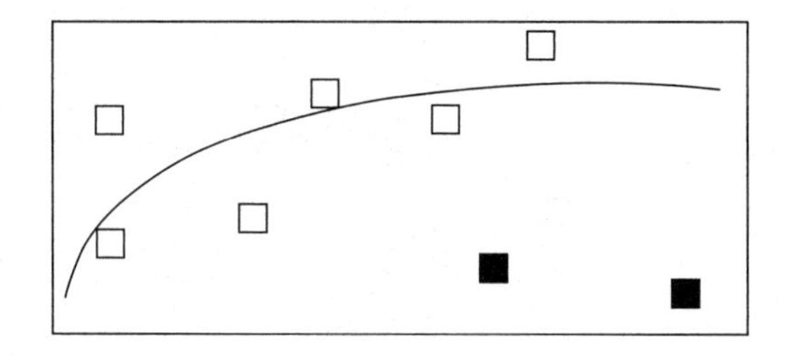

1. How should the data from the two suspected runs be handled?

2. Should the data be included in tests of statistical significance and why/why not?

3. What sources of information could Kathleen and Deborah use to help decide?

Activity

Ethics Role Play: The Sharing of Research Materials

(Modified from "On Being a Scientist: Responsible Conduct in Research," 2nd ed., National Academy Press, 1995)

The Players:

Ed is an undergraduate student majoring in biochemistry doing a senior honors thesis.

Maya, a fourth-year graduate student, is his direct mentor in the lab.

Abdul is a post-doctoral fellow, who just joined the lab.

The Situation:

Ed is frustrated because he is having trouble getting an assay to work correctly. He has tried everything he can think of and now turns to his mentor for help.

Ed: "Maya, I still can't get that assay to work reliably. Sometimes it does, but usually I get nothing. Even when I do get data, I don't know if they're any good. Can you think of anything else to try?"

Maya: "You've tried everything I can think of. When we did the experiment together, your technique was flawless, so it must be that catalyzing reagent. Have you talked to Abdul about it?"

Ed: "Abdul?"

Maya: "Oh that's right, you haven't met him yet. He's a new post-doc in the lab, who just started this week. The lab he did his PhD in used similar assays. I bet they used that catalyzing reagent, or something like it. You should ask him."

Ed: "Thanks, I'll go introduce myself."

Ed finds Abdul in Professor Lowrey's office. When he approaches, they invite him in and Professor Lowrey introduces them to one another.

Ed: "Nice to meet you, Abdul, and welcome to the lab. I'd love to talk to you about my project sometime, and learn about yours."

Abdul: "Me, too. I'm anxious to learn about all the different projects going on in the lab. Wanna' have lunch tomorrow?"

Ed: "Sounds good. I have class until 11:45. I'll come find you after that."

Abdul: "Great."

The next day at lunch, Ed describes his project, and the problems he's having with the catalyzing reagent to Abdul.

Abdul: "You should contact my old lab. They use that reagent, and have developed a form of it that yields good results. I'm sure they would send you some if you asked."

Ed: "Thanks for the tip! I'll send them an email today."

Ed writes the professor in charge of Abdul's former lab, which is at another university. She writes back and says they are still struggling to develop and characterize the reagent and are not ready to share it yet. Ed is frustrated.

Ed: "Your old mentor claims that their catalyzing reagent isn't ready for prime time, so she doesn't want to share it yet."

Abdul: "That's ridiculous. They just don't want to give you a break."

1. What kinds of information are appropriate for researchers to share with their colleagues? What about when they change labs?

2. Where can Ed go for help in obtaining materials?

3. Should Ed have contacted the professor at the other university directly himself?

4. Are there risks involving other people in this situation?

Activity

Ethics Case: Credit Where Credit is Due

(Modified from "On Being a Scientist: Responsible Conduct in Research," 2nd ed., National Academy Press, 1995)

Bea, a junior, was working on a research project that focused on developing a new experimental technique. To present her work at the Undergraduate Symposium, she prepared a poster outlining the new technique. During the poster session, Bea was surprised and pleased when Dr. Freeman, a leading researcher on campus, engaged her in a conversation. Dr. Freeman asked extensively about the new technique, and she described it fully, happy to be confidently discussing her work with a fellow scientist. Bea's faculty advisor had encouraged his students to openly share their research with other researchers, and Bea was flattered that Dr. Freeman was so interested in her work.

Six months later Bea was leafing through a journal when she noticed an article by Dr. Freeman. The article described an experiment that clearly depended on the technique that Bea had developed. She did not mind, in fact, she was somewhat flattered that her technique so strongly influenced Dr. Freeman's work. But when she turned to the citations, expecting to see a reference to her abstract or poster, her name was nowhere to be found.

1. Does Bea have any way of receiving credit for her work?

2. Should she contact Dr. Freeman in an effort to have her work recognized?

3. Is Bea's faculty advisor mistaken in encouraging his students to be open about their work?

Assignment
Due Session 21

Scientific Poster Hunt

Explore the halls in the building where your research group resides (or another research building) to find scientific posters hanging on the walls. Select one favorite, and one least favorite poster. Identify the characteristics of each poster that make it your favorite or least favorite.

1. Favorite Poster Title

   What characteristics make this poster your favorite?

2. Least Favorite Poster Title

   What characteristics make this poster your least favorite?

SESSION 21

Making Effective Scientific Presentations

GOALS

Students will

- identify and discuss the characteristics of effective research posters and oral presentations.
- recognize that scientific presentations are opportunities for discussion with colleagues about their research, not only for reporting about their research.

OUTLINE

Core Activities

Check-in

Elements of Effective Research Presentations: Results of the scientific poster hunt

Scientific Poster Web Resources

Optional Activities

Guest speaker gives scientific presentation as example (poster or short talk)

Materials for Students

- Examples of effective posters/oral presentations
- Poster/Oral Presentation Draft #1 Assignment

Possible Readings for Session 22

- None

Assignment For Session 22

- Researching Research Careers (Session 15)
- Draft #1 of poster/oral presentation (small 9" x 11") due Session 23

FACILITATOR NOTES

Core Activities

Check-in

Ask students to share whether they have ever given a research presentation. If so, ask them to describe the experience in one word.

Elements of Effective Research Presentations: Results of the Scientific Poster Hunt

Put students into pairs or small groups to share their notes about the posters they viewed in preparation for class. Have each group generate a list of the features of good scientific posters. Bring the students together as a large group to share their lists and generate a common list. The facilitator can summarize this list and distribute it to the students as a guide to use when generating their posters. Ideas that have emerged include

- eye catching, yet subtle colors
- clear, concise title
- lots of figures and/or images, not a lot of words
- large font size
- clear statement of hypothesis or research question
- diagrams/images used to explain experimental techniques

Discussion questions:

- What is/are the goal(s) of a research presentation?
- What characteristics of posters make them most effective?
- What characteristics of oral presentations make them most effective?

Scientific Poster Web Resources

Find an up-to-date list of web-based resources about how to create effective scientific presentations on the Freeman and Company's Faculty Lounge web site (www.whfreeman.com/facultylounge).

Optional Activities

Guest Speaker

Invite an experienced researcher (e.g. graduate student, post-doc, scientist) to give a research presentation at the end of class—either present a poster or give a short talk. The facilitator should lead a discussion about what information should go on the poster/slide, and what should be left to present orally. Ideas that have emerged in response to these questions in previous offerings include

- **What pieces of information or explanation did the presenter add to the information given on the poster/slide?**
  - details about the experimental design
  - personal stories about doing the research, including setbacks how they were overcome (trouble-shooting strategies)
- **How did the oral presentation of the research add to your understanding of the research poster or slides being presented?**
  - opportunity to ask questions when something on the poster was unclear
  - opportunity to discuss the connection of my own research to the research presented in the poster
  - opportunity to meet the author

Wrap-Up

Assignments for next time:

- Researching Research Careers (Session 15)
- Draft #1 of poster/oral presentation (small 9" x 11") due Session 23

Reflections and Notes

STUDENT MATERIALS

To download electronic versions of student materials go to Faculty Lounge (www.whfreeman.com/facultylounge)

Assignment
Due to peer reviewers 24 hours before Session 23

Poster/Oral Presentation Draft #1

Use your research project outline assignment to generate the first draft of a research poster or oral presentation. Use the peer review rubric and the web resources provided by the facilitator for advice on how to best construct your presentation.

Assignment
Due Session 22

Poster Peer Review Form

Reviewer: ______________________ Author: ______________________

Poster Title:

As a reviewer (in your own words), summarize what this research project is about and why it is important.

Use the rubric to evaluate each component. Offer **specific** suggestions for improvement with each rating.

Title & Authors 0 1 2 3
Comments/Suggestions:

Abstract 0 1 2 3
Comments/Suggestions:

Introduction 0 1 2 3
Comments/Suggestions:

Hypothesis or Research Question	0	1	2	3

Comments/Suggestions:

Research Methods	0	1	2	3

Comments/Suggestions:

(Expected) Results	0	1	2	3

Comments/Suggestions:

Conclusions	0	1	2	3

Comments/Suggestions:

References and Acknowledgements	0	1	2	3

Comments/Suggestions

Presentation	0	1	2	3

Comments/Suggestions

Poster Evaluation Rubric

	0	1	2	3
Title and Authors	Absent	Title is lengthy and unclear	Title is lengthy, but clear	Title is concise and clear
Abstract (brief summary)	Absent	Does not summarize the research & results. Provides only basic background information.	Summarizes research & results, but provides too much information, or uses jargon without definitions.	Short, descriptive summary of research and results; provides only relevant information in simple language.
Introduction (Background, context and relevance)	Absent	The background information presented lacks the content needed to understand the scientific basis of the hypothesis or research question.	Relevant background information is presented, but poorly organized. Therefore, the hypothesis or research question does not follow logically from it.	Relevant background information is presented and organized such that the hypothesis or research question follows logically from it.
Hypothesis or Research Question (purpose or aim/goal)	Absent	A statement is made, but it is neither a hypothesis nor a research question.	A hypothesis or research question statement is made, but it is neither concise nor follows logically from the background information.	A clear and concise hypothesis or research question statement is made that follows logically from the background information.
Research Methods (How was it done?)	Absent	Experiments are listed but lack detail and are not connected to the stated hypothesis or research question.	Experiments are listed, and either well explained or connected to the stated hypothesis or research question, but not both.	Experiments are listed, well explained, and connected to the stated hypothesis or research question.
(Expected) Results (What did you learn?)	Absent	Results are described, but lack a figure to represent them or a statement of whether they support the stated hypothesis or research question.	Results are described, but lack either a figure to represent them or a statement of whether they support the stated hypothesis or research question.	Results are presented in a figure and a statement about whether they support the stated hypothesis or research question is made.
Conclusions (What does it mean?)	Absent	Some obvious conclusions are not identified, or are not connected to the results.	The conclusions are not well connected to the results.	Clear and relevant conclusions that follow logically from the results.
References & Acknowledgements	Absent	Few or not credible references are listed.	Credible references are listed, but not cited in text.	Credible references are listed and cited in text.
Presentation (Is it visually appealing?)	Absent	Figures/Images are absent; text is too small, lengthy and detailed.	Figures are present, but are cluttered or unclear; text is large enough, but too lengthy.	Clear figures with concise text outlining the most important points.

Assignment
Due Session 22

Oral Presentation Peer Review Form

Reviewer: __________________________ Author: __________________________

Presentation Title:

As a reviewer (in your own words), summarize what this research project is about and why it is important.

Use the rubric to evaluate each component. Offer **specific** suggestions for improvement with each rating.

Title & Authors 0 1 2 3
Comments/Suggestions:

Abstract 0 1 2 3
Comments/Suggestions

Introduction 0 1 2 3
Comments/Suggestions

Hypothesis or Research Question 0 1 2 3
Comments/Suggestions

Research Methods 0 1 2 3
Comments/Suggestions

(Expected) Results 0 1 2 3
Comments/Suggestions

Conclusions 0 1 2 3
Comments/Suggestions

References and Acknowledgements 0 1 2 3
Comments/Suggestions

Presentation 0 1 2 3
Comments/Suggestions

Oral Presentation Evaluation Rubric

	0	1	2	3
Title and Authors	Absent	Title is lengthy and unclear.	Title is lengthy, but clear.	Title is concise and clear.
Overview (brief summary)	Absent	Did not summarize the research & results. Provided insufficient background information.	Summarized the research & results, but provided too much information, or used jargon without definitions.	Gave a short, descriptive summary of research and results; provided only relevant information in simple language.
Introduction (Background, context and relevance)	Absent	The background information presented lacked the content needed to understand the scientific basis of the hypothesis or research question.	Relevant background information was presented, but poorly organized. Therefore, the hypothesis or research question did not follow logically from it.	Relevant background information was presented and organized such that the hypothesis or research question followed logically from it.
Hypothesis or Research Question (purpose or aim/goal)	Absent	A statement was made, but it was neither a hypothesis nor a research question.	A hypothesis or research question statement was made, but it was neither concise nor followed logically from the background information.	A clear and concise hypothesis or research question statement was made that followed logically from the background information.
Research Methods (How was it done?)	Absent	Experiments were described but lacked detail and were not connected to the stated hypothesis or research question.	Experiments were described, and were either well explained or connected to the stated hypothesis or research question, but not both.	Experiments were well explained and connected to the stated hypothesis or research question.
(Expected) Results (What did you learn?)	Absent	Results were orally presented, but not in a figure. A statement of whether they supported the stated hypothesis or research question was not made.	Results were described, but lacked either a figure to represent them or a statement of whether they supported the stated hypothesis or research question.	Results were presented in a figure and a statement about whether they support the stated hypothesis or research question was made.
Conclusions (What does it mean?)	Absent	Some obvious conclusions were not identified, or were not connected to the results.	The conclusions were not well connected to the results.	Clear and relevant conclusions were presented and followed logically from the results.
References & Acknowledgements	Absent	Few or not credible references were listed.	Credible references were listed.	Credible references were listed and connected to the work presented.
Presentation	Absent	Speaking style was choppy, there was poor eye contact, and transitions between slides were not made.	Speaker was clear, but had poor eye contact and made only weak connections between slides.	Speaker was clear, made eye contact, and effectively transitioned between slides.

SESSION 22

Research Careers

GOALS

Students will

- explore careers for which research training in the sciences is required or recommended.
- identify how research thinking and technical skills are transferable to many science and non-science careers.

OUTLINE

Core Activities

Check-in

Career Interview Presentations

Optional Activities

Discussion of Transferability of Research Skills

The Next Step in Your Career: Factors to Consider

Materials for Students

- None

Possible Readings for Session 23

- None

Assignment for Session 23

- Presentation Draft #1—due to peer reviewers 24 hours before Session 23
- Peer Reviews—due Session 23

FACILITATOR NOTES

Core Activities

Check-in

Ask students to share what type of career they explored.

Research Career Interview Presentations

Give students a few minutes to reflect on the discussion questions before they begin presenting their interviews to the group. Allow each student 5–10 minutes to present the interview and to answer questions about the career he explored. Discussion questions:

- Why did you explore your chosen career?
- What, if any, preconceived ideas about the career have changed based on what you learned?
- Why is research training important for the career you explored?

Optional Activities

Research Skill Transferability Discussion

The goal of this discussion is to get students to reflect on the skills they are gaining through their research experience, including how those skills may be useful in non-research careers.

Discussion questions:

- **What specific research thinking and technical skills are required for the career you explored?**
    - critical thinking and problem solving
    - specific technical skills for careers that involve doing research day-to-day

- **In what other careers, including non-science careers, might these skills be useful or required?**
  - teaching
  - leadership, management, decision-making positions
- **Do you think that learning to do research should be a required part of the undergraduate curriculum in your discipline? Why or why not?**
  - Yes—it is the best way to understand the discipline; everyone needs an appreciation for how new knowledge is generated.
  - No—not everyone needs these skills to get a job; it is not practical to require this of everyone; lab classes are enough.

The Next Step in Your Career: Factors to Consider

The goal of this exercise is to help students think about what they need to do to prepare for the next stage of their academic or professional careers, and to expose them to the kinds of decisions they will need to make. Allow students 5–10 minutes to cut out and individually rank the factors. Then have them pair up and explain their rankings to their partners. Finally, survey the group to identify and discuss the most common top priorities. Ideas that have emerged in discussion of this question in previous offerings include

- **How can you learn about or investigate the factors that are most important to you?**
  - Find a mentor in the field to guide you
  - Do research on-line
  - Talk to current students in the programs of interest
  - Ask the graduate training or professional program administrator

Wrap-Up

Assignments for next time:

- Poster Draft #1—due to peer reviewers 24 hours before Session 23
- Peer Reviews—due Session 23

Reflections and Notes

STUDENT MATERIALS

To download electronic versions of student materials go to Faculty Lounge (www.whfreeman.com/facultylounge)

Activity

The Next Step in Your Career

Factors to Consider

Each box lists a factor that you may want to consider when selecting a graduate or professional school program, or a job. There are also two blank boxes in which you may add factors that are important to you that are missing from the list. Cut out the boxes and rank the factors in order of importance to you.

Opportunity to work with a specific advisor, mentor, physician, or teacher	Coursework requirements
Climate of the training environment	Location
Relative value of teaching and research training	Alignment of personal goals with the offerings/ opportunities
Reputation of a specific advisor, mentor or co-worker	Funding
Reputation of the department, office, program or institution	Happiness of other graduate/medical students/ co-workers in the program
Range of academic opportunities to engage beyond graduate study	Feeling of inclusively—seeing that there are others like you there
Type of preliminary/qualifying exam	Type of curriculum (e.g. case-based, traditional lecture, clinical)

SESSION 23

Presentation Peer Review Draft #1

GOALS

Students will

- ▶ be able to give constructive feedback on their peers' presentations.

OUTLINE

Core Activities

Check-in

Peer Feedback Discussion

Optional Activities

Discussion of Revision Plans

Materials for Students

- None

Possible Readings for Session 24

- None

Assignment for Session 24

- Presentation Draft #2—due Session 24

To Do for Session 24

- **Invite "outside" reviewers to attend Session 24.** These can be graduate students, post-docs, scientists, and/or faculty; anyone with experience creating and presenting scientific presentations. Share a copy of the peer review rubric with them before the session, so they are familiar with the expectations.

FACILITATOR NOTES

Core Activities

Check-in

Ask students to identify one thing they would like to improve in their presentation draft. Ask reviewers to address these items in their feedback.

Poster Peer Review

Students should pair up (or get into groups) with their peer review partners. Alternatively, if the group is small enough, poster or talk drafts may be projected and reviewed by everyone. Large group review is feasible only if there is enough time to get to everyone in one class period (or if another class period is added).

Before students exchange reviews, offer a few moments of silence to identify

- the best part of the presentation,
- one specific suggestion for how to improve the presentation, and
- how they would suggest improving the one thing the author identified during check-in.

Optional Activities

Revision Plans Discussion

The goal of this discussion is to give students the opportunity to process the feedback they received and to identify specific strategies they will use to improve their presentation.

Discussion questions:

- What are the major issues you need to address in your presentation?
- Specifically, how will you address these issues in your second draft? To whom, if anyone, will you go for help in revising?

Wrap-Up

Assignments for next time:

- Presentation Draft #2—due Session 24

Reflections and Notes

STUDENT MATERIALS

To download electronic versions of student materials go to Faculty Lounge (www.whfreeman.com/facultylounge)

SESSION 24

Presentation Outside Review Draft #2

GOALS

Students will

- practice presenting their research presentations to a new audience.
- receive feedback on their presentation from an outside reviewer.

OUTLINE

Core Activities

Check-in

Outside Review of Presentation Draft #2

Optional Activities

Discussion of Revision Plans

Materials for Students

- None

Possible Readings for Session 25

- None

Assignment for Session 25

- Final Presentation

FACILITATOR NOTES

Core Activities

Check-in

Ask the students and outside reviewers to introduce themselves. Students should share the area of their research and the lab in which they work. Outside reviewers should share their position (e.g. graduate student, post-doc, professor), their department, and their research area of expertise.

Outside Reviews of Poster Draft #2

Students should sit around the room, at individual tables with mini-copies (8½" × 11") of their presentation in front of them. Reviewers rotate around the room to visit each student individually (or in pairs). Students present a fresh mini-copy of their presentation to each reviewer, on which they can make notes about the feedback they receive. At the end of the session, students will have several copies of their presentation with specific feedback from each reviewer.

Optional Activities

Discussion of Revision Plans

The goal of this discussion is to give students the opportunity to process the feedback they receive and to identify specific strategies they will use to improve their presentation.

Discussion questions:

- What are the major issues you need to address in your presentation?
- Specifically, how will you address these issues in your final draft? To whom, if anyone, will you go for help in revising?

Wrap-Up

Assignments for next time:

- Final Presentation

Reflections and Notes

STUDENT MATERIALS

To download electronic versions of student materials go to Faculty Lounge (www.whfreeman.com/facultylounge)

CHAPTER 25

Final Presentations

GOALS

Students will

- present final posters or research talks.

OUTLINE

Core Activities

Presentations

Optional Activities

Presentation Evaluation Exercise

Materials for Students

- Presentation Review Rubric (Session 21)
- Research Group Funding Assignment

Possible Readings for Session 26

- None

Assignment for Session 26

- Research Group Funding

To Do for Session 26

- Invite someone from your campus internships/fellowships office to present at Session 26.

FACILITATOR NOTES

Core Activities

Poster Presentation

If your campus sponsors a formal event at which undergraduate students can share their independent research and/or creative works, then this would be an ideal venue in which to have the students present their posters. Alternatively (or in addition), the facilitator(s) can organize a mini-symposium for the students and invite faculty, scientists, post-docs, graduate, and undergraduate students to attend.

Optional Activities

New Presentation Evaluation Exercise

If there are multiple sections of Entering Research being offered on your campus, then students can be assigned to visit the poster/presentation of a peer who attends another section. In addition to the informal conversation that the students will have about the work, they can complete a review rubric (Session 21) about the presentation as an assignment. The reviews can then be shared with the presenter.

If the presentations are part of a larger symposium, students can also be assigned to review presentations given by students not in the course.

Wrap-Up

Thank students and guests for participating in the symposium.

Assignments for next time:

- Research Group Funding

Reflections and Notes

STUDENT MATERIALS

To download electronic versions of student materials go to Faculty Lounge (www.whfreeman.com/facultylounge)

Assignment
Due Session 26

Research Group Funding

Meet with the principal investigator of your research group and ask him/her how the research in your group is funded. Specifically ask

- What are the types of expenses in your research group (e.g. supplies, animals, salaries)?
- From where do the funds for these expenses come (e.g. federal or private grants, the state, the university or college)?
- Who is responsible for securing the research funds and how much time does this person(s) spend doing it?

SESSION 26

The Future of Your Project

Funding/Grants

GOALS

Students will

- learn how research is funded.
- define the next steps in their research project.

OUTLINE

Core Activities

Check-in

Discussion of Research Funding

Funding Opportunities Presentation

Optional Activities

Discussion of Mini-Grant Assignment: The future of your project

Materials for Students

- Mini-grant proposal assignment

Possible Readings for Session 27

- Slaughter, G. R. (2006) Chapter 8—**Writing Research Proposals**, *Beyond the Beakers: Smart Advice for Entering Graduate Programs in the Sciences and Engineering*, Baylor College of Medicine. www.bcm.edu/gs/BeyondTheBeakers/Table%20of%20Contents.htm

Assignment for Session 27

- Mini-grant proposal draft
- Discussion Questions
  - What advice would you give to new students interested in doing a research experience?

- How has your research experience prepared you for your future career? Identify the skills you have learned that will be useful in your future career and explain how?
- Outline at least three differences between your first few months in lab and where you are now.

FACILITATOR NOTES

Core Activities

Check-in

Ask students to reflect on their poster/oral presentation experience. How did it go?

Discussion of Research Funding

The goal of this discussion is to learn about the expenses associated with doing research, how these expenses are funded, and who is responsible for securing those funds. Students can share, either in small groups or as a large group, what they learned from the principal investigators about their own research groups' funding. Ideally a diversity of funding situations will be presented from which the students can learn. Ideas that have emerged in response to these questions in previous offerings include:

- ▶ **How is the research in your group funded?**
    - federal grants
    - grants from private industry
    - university or college funds
    - individual student fellowships
    - charitable foundation
- ▶ **Why is it important for scientists to be able to write compelling research proposals?**
    - this is the primary mechanism by which research is funded
    - it helps to define the research project for everyone involved in it
    - it is helpful when writing final reports of research
- ▶ **How can you, as an undergraduate, begin to develop the skills needed to write research proposals?**
    - practice writing proposals for undergraduate research fellowships
    - take a science writing class

- **What opportunities are available for undergraduates to fund their research?**
  - college/university research fellowships or internships
  - professional society fellowships
  - supplements to mentor's research grants

Funding Opportunities Presentation

The facilitator may lead a discussion about undergraduate research funding opportunities on their campus (and beyond), or, if the campus has an undergraduate internships/fellowships office, a person from that office can give a presentation.

Optional Activities

Discussion of Mini-Grant Assignment

What are the next steps in your research? Facilitators can present this last assignment as an opportunity for students to outline what they would propose in a grant to fund the next phase of their research project (e.g. if they were going to continue it over the summer, or into the next semester).

Wrap-Up

Assignments for next time:

- Mini-grant draft proposal

Reflections and Notes

STUDENT MATERIALS

To download electronic versions of student materials go to Faculty Lounge (www.whfreeman.com/facultylounge)

Assignment
Due to peer reviewers 24 hours before Session 28

Mini-Grant Proposal

The Next Step in Your Research Project

The goal of your mini-grant proposal is to convince fellow scientists (peer reviewers) that your research is worth funding. It is meant to be a continuation of the research you have been doing (the next step), not a new project. Writing it is an opportunity to **work with your mentor** to pull together the pieces you've written so far, plan the future (next step) of your project, and create a proposal draft that could be revised and submitted for funding later.

The proposal should be **3–5 pages (double spaced)**, including the five sections outlined below, and an appendix with a budget.

1. Abstract
A brief summary of the proposed research (200 words or less).

2. Introduction: Background and Relevance
A brief description of the background and importance of the proposed research.

3. Hypothesis/Research Question OR Specific Aims
EITHER a statement of the hypothesis/research question, OR a list of 1–3 specific objectives.

4. Preliminary Results
A brief presentation of preliminary data that is relevant to the proposed research project. This could be data collected by you or others in the research group, or found in the published literature.

5. Proposed Experiments with Timeline
A detailed description of the proposed experiments, including methods and a timeline for completion.

Appendix: Budget

An estimate of the supply costs (in dollars) and personnel time (in hours) needed to complete the proposed research project. Personnel time should include both student and mentor time. It is important to work with your mentor on the budget to ensure your estimates are realistic.

Assignment
Due Session 28

Peer Review of Mini-Grant Proposal

Author ________________________ Reviewer ________________________

Strengths:

Weaknesses:

Overall Score (circle)

1 **Excellent**—Outstanding proposal in all respects, deserves highest priority for support.
2 **Very Good**—High quality proposal in nearly all respects, should be supported if at all possible.
3 **Good**—A quality proposal, worthy of support.
4 **Fair**—Proposal lacking in one or more critical aspects, key issues need to be addressed.
5 **Poor**—Proposal has serious deficiencies.

SESSION 27

Mini-Grant Proposal Peer Review

GOALS

Students will

- receive feedback from peers on their mini-grant proposal draft

OUTLINE

Core Activities

Check-in

Peer Feedback Discussion

Optional Activities

None

Materials for Students

None

Possible Readings for Session 28

- None

Assignment for Session 28

- Final mini-grant proposal

FACILITATOR NOTES

Core Activities

Check-in

Do a round robin and ask students to reflect on how easy/difficult it was to review their peers' work. Did they know what to look for when reviewing, now that they have done this several times?

Mini-Grant Proposal Peer Review

Students should pair up (or get into groups) with their peer review partners. Alternatively, if the group is small enough, all drafts may be discussed as a group. This is feasible only if there is enough time to get to everyone in one class period.

Before students exchange reviews, offer a few moments of silence to identify

1. the best part of the mini-grant proposal, and
2. one specific suggestion for how to improve the mini-grant proposal.

Wrap-Up

Assignments for next time:

- Final mini-grant proposal

Reflections and Notes

STUDENT MATERIALS

To download electronic versions of student materials go to Faculty Lounge (www.whfreeman.com/facultylounge)

SESSION 28

Research Experience Reflections and Celebration

GOALS

Students will

- reflect on their research experience and their participation in the workshop.
- celebrate the research they've completed so far, and discuss future plans.

OUTLINE

Core Activities

Check-in

Entering Research Part II Research Experience Reflections

Discussion of Research Experience

Optional Activities

Post-survey (www.whfreeman.com/facultylounge)

FOOD!

Materials for Students

- Entering Research Part II Workshop Evaluation

FACILITATOR NOTES

Core Activities

Check-in

Ask students to share the most important thing they learned in their research experience.

Entering Research Part II Research Experience Reflections

Give students 10–15 minutes to complete the Research Experience Reflections worksheet.

Discussion of Research Experience (so far)

Ask students to share their research experience reflections. Either the facilitator or a student could facilitate this discussion. Reflections that previous groups have contributed are listed below.

- **What advice would you give to new students interested in doing a research experience?**
  - Be persistent when looking for a mentor, but not pushy.
  - Don't be afraid to ask your mentor questions.
  - Be creative with your scheduling to allow blocks of time for research
  - Go to lab meetings, conferences
  - Find a lab that is doing research that interests you
- **How has your research experience prepared you for your future career? Identify the skills you have learned, that will be useful in your future career and explain how?**
  - Problem-solving or trouble-shooting skills
  - Critical-thinking skills
  - Writing and presentation skills
  - The importance of doing something right, paying attention to details
  - Knowing general career options

► **Outline at least three differences between your first few months in lab and where you are now.**
- Didn't know anything, now feel more comfortable
- Felt in the way, but can contribute to research group now
- Now we can design our own experiments

Entering Research Part II Workshop Evaluation

Give students 10–15 minutes to complete the workshop evaluation.

Optional Activities

Post-Survey

If you are participating in the research study, ask students to complete the online survey, the link to which can be found on Freeman and Company's Faculty Lounge web site (www.whfreeman.com/facultylounge).

Wrap-Up

Invite students to share parting thoughts, concerns, or questions.

Collect Research Experience Reflections worksheets and student evaluations.

Will the group keep in touch in the future? If so, how?

Reflections and Notes

STUDENT MATERIALS

To download electronic versions of student materials go to Faculty Lounge (www.whfreeman.com/facultylounge)

Entering Research Part II

Reflections on Your Research Experience

1. Was your research experience what you expected it to be? Why or why not?

2. What academic and personal goals did you achieve in your research experience? How do they compare to the goals you outlined at the beginning of your experience?

3. What values, experiences, and/or perspectives did you contribute to your research team? Were you able to contribute in ways that you did not predict? How?

4. How did you overcome your greatest concern about doing research? What was the most challenging aspect of your research experience?

5. What was the best part about your research experience? Are you planning to continue doing research? Why or why not?

Entering Research Part II

Evaluation

1. How helpful were these sessions to your learning about and engaging in independent research?

	Do Not Remember	Not Helpful	Somewhat Helpful	Helpful	Very Helpful
Workshop Intro. & Science Comm.	○	○	○	○	○
Project Outline & Science Abstract	○	○	○	○	○
Science & Society	○	○	○	○	○
Peer Review Public Abstract	○	○	○	○	○
Research Ethics	○	○	○	○	○
Making Effective Presentations	○	○	○	○	○
Research Careers	○	○	○	○	○
Peer Review Draft #1	○	○	○	○	○
Outside Review Draft #2	○	○	○	○	○
Final Presentations	○	○	○	○	○
Funding/Grants—Project Future	○	○	○	○	○
Reflections & Celebration	○	○	○	○	○

Please comment on any of the above sessions.

2. How helpful were these assignments in supporting your independent research experience?

	Do Not Remember	Not Helpful	Somewhat Helpful	Helpful	Very Helpful
Project Outline & Science Abstract	○	○	○	○	○
Reflect Mentor Relationship	○	○	○	○	○
Project Outline Peer Feedback	○	○	○	○	○
General Public Abstract	○	○	○	○	○
General Abstract Peer Review	○	○	○	○	○
Researching Research Careers	○	○	○	○	○
Ethics Case Disc. with Mentor	○	○	○	○	○
Scientific Poster Hunt	○	○	○	○	○
Presentation Peer Review	○	○	○	○	○
Presentation Outside Review	○	○	○	○	○
Final Presentation Peer Review	○	○	○	○	○
Your Research Group's Funding	○	○	○	○	○
Mini-grant Proposal	○	○	○	○	○
Research Exp. Reflections	○	○	○	○	○

Please comment on any of the above assignments.

3. How effective were each of these class formats used in the workshop series?

	Do Not Remember	Not Effective	Somewhat Effective	Effective	Very Effective
In-class Discussions	○	○	○	○	○
On-line Discussions	○	○	○	○	○
Case Studies	○	○	○	○	○
Role Play Exercises	○	○	○	○	○
In-class Activities	○	○	○	○	○
Student Disc. Facilitation	○	○	○	○	○
Instructor Disc. Facilitation	○	○	○	○	○

Please comment on any of the above class formats.

4. Any additional comments about the workshop series?

APPENDIX

APPENDIX 1

Facilitating, Not Teaching

Some Practical Tips

To be a facilitator is to

- act as a guide on the side (not to teach or lecture about the content);
- lead, but not dominate the discussion;
- intervene when destructive behavior occurs;
- be flexible enough to let the group members' experiences drive the discussion;
- allow silence when group members need time to think;
- ensure that everyone has a chance to contribute to the discussion; and
- summarize and bring closure to the discussion.

Facilitating is a skill that can be practiced and improved. For those who have a lot of experience lecturing in the classroom, the transition to facilitating discussions can be challenging. This is true for the facilitator and the students, both of whom are comfortable with the traditional classroom teacher and student roles. To aid this transition, here are several strategies designed to help new facilitators deal with common challenges.

What do I do when no one talks?

- Have everyone write an idea, thought, or answer to a question on a slip of paper and toss it in the middle of the table. Each participant then draws a slip of paper from the center of the table (excluding their own) and reads it out loud. All ideas are read out loud before any open discussion begins.

- Have participants discuss a topic in pairs for 3–5 minutes before opening the discussion to the larger group.

What do I do when one person is dominating the conversation?

- Use a "talking stone" to guide the discussion. Participants may talk only when holding the stone. Each person in the group is given a chance to speak before anyone else can have a second turn with the stone. Participants may pass if they choose not to talk. Importantly, each person holding the stone should express his or her own ideas and resist from responding to someone's ideas. Generally once everyone has a chance to speak, the group can move into open discussion without the stone.
- Ask the dominating student to stay after class to discuss how she can help you get the quiet students to participate in the discussion. Mention that she is an excellent discussion participant and that you would like her to ask her peers more questions to get them involved.

What do I do when a certain person never talks?

- Each week have a different participant lead off the discussion so that every person in the group has the chance to speak first during the course of the seminar.
- Assign participants in the group different roles in a scenario or case study and ask them to consider the case from a certain perspective. Ask the participants to discuss the case in the larger group from the various perspectives. For example, some participants could consider the perspective of the mentor, while others consider the perspective of the mentee.
- Try some smaller group discussions (2–3 participants per group) as the person may feel more comfortable talking in the smaller group.

What do I do when the group members direct all their questions and comments to me, instead of their fellow group members?

- Each time a group member talks to you, move your eye contact to someone else in the group to help the speaker direct his attention elsewhere.

- Ask the participants for help in resolving one of your own mentoring challenges. For example, ask them for advice on how to deal with a graduate student who does not get along with the research group technician. If you implement the idea, tell the participants how it went. This helps the group members stop looking to you for the right answers.

What do I do when the group gets off target?

- Have everyone write the ideas they want to share on the topic for 3 minutes. This short writing time will help participants collect their ideas and decide what thoughts they would most like to share with the group so they can focus on that point.
- Ask someone to take notes and recap the discussion at the half-way and end points of the session to keep the conversation focused.

What do I do when a group gets too comfortable; when they refuse to disagree with one another and the conversation goes nowhere?

- Use a role-play scenario and assign group members to play different, opposing roles.
- Split the group in two and organize a debate about a controversial issue that is relevant to the day's topic. Ask each group to develop an argument for one side of the issue, and then present and defend their arguments.

What do I do when the group members disagree in ways that are disrespectful; when there is "open warfare?"

- Model what respectful discourse looks like by stopping the conversation and restating the disagreeing statements in a respectful way.
- If the statements are personal, broaden the conversation by pointing out how the statement(s) made can (or cannot) be generalized. Describe a hypothetical situation that is relevant to the topic, but not specific to any one member of the group.
- Ask the person(s) who is/are making disrespectful comments to talk to you one-on-one after the workshop. This allows you to avoid putting them on the spot in front of others. They may not realize that their comments were disrespectful, and would benefit from knowing it.

What do I do if a group member stops attending my workshop series?

- Contact the person to
  - ask why she have been missing meetings;
  - ask if she plans to continue attending in the future; and
  - let her know that the group misses her perspective when she is absent.
- Ask one of the other group members to act as a representative of the group and to follow up with the missing person. He can specifically invite her to rejoin.

APPENDIX 2

Discussion Guidelines for Students

Role of student as FACILITATOR

As the facilitator of a discussion, you should

- be prepared to summarize the main points from the pre-class online discussion (if there was one);
- design an activity or strategy to engage classmates in the face-to-face discussion (see below for ideas);
- set a positive tone;
- keep the group focused;
- keep track of time;
- ensure that everyone is heard;
- ask questions;
- listen well; restate, clarify, and connect ideas;
- mediate conflict; and
- monitor ground rule compliance.

Facilitation Ideas/Strategies

- **Free Wheeling**—everyone contributes their thoughts randomly

- **Round Robin**—go around the circle and each group member offers his/her thoughts
- **Sliding Groups**—break out in smaller groups with specific questions to discuss or an assignment to do, then reconvene as a large group to share
- **Walk and Talk** in pairs, then reconvene to discuss
- **Slip Method**—each group member writes his/her thoughts on a piece of paper and tosses the paper into the center of the group. The group members then take someone else's paper and present the idea on it.

Role of student as RECORDER

As the recorder of a discussion, students should

- concentrate on taking notes on the main ideas of the discussion;
- write a comprehensive summary of the discussion; and
- upload the summary within 48 hours after class and name the file using the month and date the discussion took place (e.g. summary0916.doc).

Role of student as PARTICIPANT

As a participant in the discussion, students should

- be prepared for the discussion;
- actively participate;
- offer insightful and on-topic comments and questions;
- attentively and respectfully listen to peers;
- be polite when disagreeing with peers and clear about their points; and
- follow the group's discussion ground rules.

APPENDIX 3

Entering Research Part I Sample Syllabus

Instructor

NAME: EMAIL:

DEPT and OFFICE ADDRESS: PHONE:

Seminar Description

This one-credit seminar course for undergraduate students is the first in a series, designed to complement the independent research experience. It is to be taken concurrently with research credits. Students meet weekly to share their research experiences and to get feedback on the progress of their research projects.

Student Learning Objectives

The main objectives of the seminar are to help students find a research mentor, write a research project proposal, and begin doing research. Specific learning objectives in the two part series Entering Research series include

Research Process Skills

Students will

- define a research question, design question, or hypothesis for their project.
- find and evaluate relevant primary literature and background information related to their project.
- design experiments to test their hypothesis.
- learn the techniques needed to do their experiments.
- learn and follow appropriate protocols for documenting their research.
- analyze their experimental data.
- use logic and evidence to build arguments and draw conclusions about their data.
- define future research questions.

Communication

Students will

- explain the focus of their group's research, how individual research group members and projects are connected, and how the research in their group contributes new knowledge in their discipline.
- connect their research to issues relevant to society at large.
- effectively communicate their research findings in oral and written scientific formats.
- connect their research experience to what they have learned in courses.

Professional Development

Students will

- establish and maintain a positive relationship with their mentor by agreeing on common goals and expectations for the research experience, and revisit those goals and expectations regularly.
- define their roles and responsibilities as a member of their research group.
- define and contribute to discussions about the forms and consequences of scientific misconduct.
- contribute to peer review in the learning community and explain the role of peer review in science.
- know the mechanisms for funding research.
- identify research career options in their discipline.

Grading

Attendance is required. Please notify your instructor(s) as soon as possible if you cannot attend due to sudden illness, family emergencies, and so forth.

10%—Attendance

10%—Pre-Class Discussion Questions

10%—In-Class Participation

25%—Assignments

10%—Hypothesis or Research Questions

15%—Experimental Design and Potential Results

20%—Research Proposal

Dates	Topics	Assignments Due
Session 0 (Before start)	Finding a Research Mentor	**PRE-survey** • Identify potential mentors
Session 1	Introduction to the Workshop Series and Finding a Research Mentor	• Contact 5 potential mentors
Session 2	The Nature of Science	• Research Experience Expectations
Session 3	Searching the Literature for Scientific Articles	• Research Topic and Key Words
Session 4	Reading Scientific Articles and Mentoring Styles	• Scientific Article Critique
Session 5	Your Research Group's Focus	• Your Research Group's Focus
Session 6	Establishing Goals and Expectations with Your Mentor	• Mentor Biography • Mentor-Mentee Contract
Session 7	Who's Who in Your Research Group	• Research Group Diagram
Session 8	Documenting Your Research	• Your Group's Research Documentation Protocol
Session 9	Defining Your Hypothesis or Research Question	• Visiting Peer Research Group • Background Information and Hypothesis or Research Question
Session 10	Designing Your Experiments	• Experimental Design & Potential Results with Timeline
Session 11	Research Proposal Review Draft #1	• Research Proposal Draft #1 • Peer Reviews
Session 12	Research Proposal Review Draft #2	• Research Proposal Draft #2 • Peer Reviews
Session 13	Final Research Proposal Presentations	• Final Research Proposal
Session 14	Final Research Proposal Presentations (continued)	• Research Experience Reflections • **POST-Survey and Workshop Evaluation**

APPENDIX 4

Entering Research Part II Sample Syllabus

Instructor

NAME: EMAIL:

DEPT and OFFICE ADDRESS: PHONE:

Seminar Description

This one-credit seminar course for undergraduate students is the second in a series, designed to complement the independent research experience. It is to be taken concurrently with research credits. Students meet weekly to share their research experiences and to get feedback on the progress of their research projects.

Student Learning Objectives

The main objectives of the seminar are to support students as they complete their research projects, present their findings in a public venue, and write a mini-grant proposal for funding the next phase of their project. Specific learning objectives in the two part series Entering Research series include

Research Process Skills

Students will

- define a research question, design question, or hypothesis for their project.
- find and evaluate relevant primary literature and background information related to their project.
- design experiments to test their hypothesis.
- learn the techniques needed to do their experiments.
- learn and follow appropriate protocols for documenting their research.
- analyze their experimental data.
- use logic and evidence to build arguments and draw conclusions about their data.
- define future research questions.

Communication

Students will

- explain the focus of their group's research, how individual research group members and projects are connected, and how the research in their group contributes new knowledge in their discipline.
- connect their research to issues relevant to society at large.
- effectively communicate their research findings in oral and written scientific formats.
- connect their research experience to what they have learned in courses.

Professional Development

Students will

- establish and maintain a positive relationship with their mentor by agreeing on common goals and expectations for the research experience, and revisit those goals and expectations regularly.
- define their roles and responsibilities as a member of their research group.
- define and contribute to discussions about the forms and consequences of scientific misconduct.
- contribute to peer review in the learning community and explain the role of peer review in science.
- know the mechanisms for funding research.
- identify research career options in their discipline.

Grading

Attendance is required. Please notify your instructor(s) as soon as possible if you cannot attend due to sudden illness, family emergencies, and so forth.

10%—Attendance

10%—In-Class Discussion Participation

10%—Project Outline and Scientific Abstract

5%—General Public Abstract

20%—Peer Reviews (10% each for Abstract and Poster)

25%—Presentation (Poster or Talk)

20%—Mini-Grant Proposal

Dates	Topics	Assignments Due
Session 15	Introduction to the Workshop Series and Science Communication	**PRE-survey**
Session 16	Research Project Outlines and Scientific Abstracts	· Research Project Outline & Science Abstract
Session 17	Research Project Outlines and Scientific Abstracts (continued)	· Reflecting on Your Mentoring Relationship
Session 18	Science and Society	· None
Session 19	Peer Review of General Public Abstracts	· Draft General Public Abstract · Peer Reviews
Session 20	Research Ethics	· Final General Public Abstract · Ethics Case Discussion with Mentor
Session 21	Making Effective Scientific Presentations	· Scientific Poster Hunt
Session 22	Research Careers	· Researching Research Careers
Session 23	Presentation Peer Review Draft #1	· Presentation Draft #1 · Peer Reviews
Session 24	Presentation Outside Review Draft #2	· Presentation Draft #2
Session 25	Final Presentations	· Final Presentation
Session 26	The Future of Your Project—Funding/Grants	· Research Group Funding
Session 27	Mini-Grant Proposal Peer Review	· Draft of Mini-Grant Proposal
Session 28	Research Experience Reflections and Celebration	· Final Mini-Grant Proposal · Research Experience Reflections **POST-Survey & Workshop Evaluation**

APPENDIX 5

Sample 10-Week Summer Syllabus

Dates	Topics	Assignments Due
Week 1 **Session 5**	**Your Research Group's Focus**	· Research Experience Expectations (Session 2) · Your Research Group's Focus
Week 2 **Session 6**	**Establishing Goals and Expectations with Your Mentor**	· Mentor Biography · Summary of Expectations
Week 3 **Session 7**	**Who's Who in Your Research Group**	· Research Group Diagram · Visiting Peer Research Group
Week 4 **Sessions 9 & 10**	**Defining Your Hypothesis or Research Question & Designing Your Experiments**	· Background Information & Hypothesis or Research Question · Experimental Design & Potential Results with Timeline
Week 5 **Session 8**	**Documenting Your Research**	· Your Group's Research Documentation Protocol
Week 6 **Session 19**	**Research Ethics**	· Ethics Case Discussion with Mentor
Week 7 **Session 20**	**Making Effective Scientific Presentations**	· Scientific Poster Hunt
Week 8 **Session 21**	**Research Careers**	· Researching Research Careers
Week 9 **Session 22**	**Peer Review of Presentations**	· Presentation Draft · Peer Reviews
Week 10 **Session 24**	**Final Research Symposium**	· Final Presentation · Research Experience Reflections

APPENDIX 6

Syllabi for Entering Research and Entering Mentoring in Parallel

Entering Research Part I	Entering Mentoring
Topics	Topics
0 Finding a Research Mentor	No Meeting
1 Introduction to Entering Research Part I and Finding a Research Experience	No Meeting
2 The Nature of Science	No Meeting
3 Searching the Literature for Scientific Articles	No Meeting
4 Reading Scientific Articles and Mentoring Styles	1 Getting Started
5 Your Research Group's Focus	2 Learning to Communicate
6 Establishing Goals and Expectations with Your Mentor	3 Establishing Goals and Expectations
7 Who's Who in Your Research Group	4 Determining Understanding
8 Documenting Your Research	5 Fostering Independence
9 Defining Your Hypothesis or Research Question	6 Addressing Diversity
10 Designing Your Experiments	7 Teaching Ethics
11 Research Proposal Review Draft #1	8 Developing a Mentoring Philosophy
12 Research Proposal Review Draft #2	No Meeting
13 Final Research Proposal Presentations	No Meeting
14 Final Research Proposal Presentations	No Meeting

J. Handelsman, C. Pfund, S.M. Lauffer, C. Pribbenow Entering Mentoring: A Seminar to Train a New Generation of Scientists, University of Wisconsin Press, Madison, 2005.

APPENDIX 7

Sample One-Day Workshop Agenda

Getting the Most from Your Research Experience

FORMAT

- Audience: Students in the early stages of their research experience (first semester).
- Research Mentors are invited to have lunch with their students and to draft a mentor–mentee contract.
- Workshop activities are done in small groups of four to five students.

AGENDA

8:15 Arrival

8:30 Welcome

- Introduce facilitator
- General overview of the workshop

8:35 Introductions, Establish/Meet Working Groups, and Establish Discussion Ground Rules

- Mix up the students so they are sitting with people they do not know.
- Name tent introductions
  - Students write their name on the front of the tent and in the four corners: hometown, favorite food, hobby, # people in family.
- Ground Rules (Session 1)
  - What do you like best and least about participating in small group discussions? OR What were the characteristics of the best and worst discussions in which you've participated?
  - Review general list; add others that students come up with.
  - Rotating facilitation – each student in the group take a card with a number from 1 to 5 on it.

9:00 Workshop Goals

Participants will

- define, or revisit, their research experience goals and expectations.
- identify and prioritize roles they expect their research mentor to play.
- identify other possible mentors.
- consider the relationships in their research group and how they can most effectively contribute to and learn from the members of this group.
- reflect on the alignment of their research experience goals and expectations with those their mentor has for them.
- contribute to defining the elements of a good research project and evaluate their own project in light of these elements.
- consider and begin to plan for the next stage in their career.

9:15 Case Study: Frustrated (Session 7)

- Work in groups to discuss case.
- Bring large group together to share.
- How would it have helped Jamal if he had established common goals and expectations with his mentors? Could this situation have been avoided?

9:45 Research Experience Goals and Expectations (Session 1)

- Work independently (10–15 minutes) to articulate your research experience goals and expectations.
- Share ideas with small group.
- Share common ideas with the large group.

10:45 BREAK

11:00 Three Professors Activity (Session 4)

- What style mentor works best for you now? Will this (or has this) change(d) with time?
- Share in small groups.

11:30 Prioritizing Research Mentor Roles (Session 17)

- Individually rank the roles you hope your research mentor will play.
- Share with a partner and explain your top two or three.
- Create a large group summary of top two roles.
- Do all of these roles need to be fulfilled by ONE (research) mentor? Who else in your research group, or beyond, could be a mentor for you?

12:00 WORKING LUNCH with Mentors

- Share goals and expectations and roles for your research mentor activities.
- Draft a Mentor–Mentee Contract (Session 1).

1:15 Debrief Discussion with Mentor—Goals and Expectations Revisions

- How did the discussion with your mentor go?
- Were your goals and expectations aligned? If not, were you able to resolve your differences?
- Were you able to draft a contract?
- How can you use this contract to keep your experience "on track?"

1:45 Elements of a Good Research Project (Sessions 9 and 10)

- Half of the small groups brainstorm the elements of a good hypothesis/research question, and the other half the elements of a good experimental design.
- Each set of groups come together to generate a common list.
- Share with the large group—create take home "master list."

Research Project Outline Worksheet (Session 15)—take-home exercise

Define (or refine) your research project using this worksheet as a guide. Consider your project in light of the "good elements master list" we generated together. Can you improve your project design?

2:45 BREAK

3:00 The Next Step in Your Career—Factors to Consider (Session 22)

Selecting the Right Graduate/Professional Program or Job

- Prioritize factors individually, ranking at least five.
- Share lists with a partner and discuss.
- Follow up large group discussion:
  - How can you learn about or investigate the factors that are most important to you as you consider which graduate or professional programs you will apply to?

3:30 Preparing for the Interview

- Give each student two index cards on which to answer these questions:
  - Index card #1—What questions would you ask in an interview?
  - Index card #2—What information about yourself do you want to make sure gets across during the interview? How will you tell them this?
- Compile index cards and discuss.

4:00 Letter of Recommendation (Session 6)

What can you do to insure a good letter of recommendation from your mentor?

- Individually complete the letter of recommendation outline.
- Share answers with small group.

Challenge: Share this with your mentor . . . when you're ready!

4:30 Wrap-Up and Evaluation

- Recap workshop goals.
- Challenge students to take ownership of their research experience.
- Any final questions or comments?
- Hand out evaluation.

Materials

- Nametags
- LCD projector and computer for facilitator
- Whiteboard, chalkboard, or flip chart for facilitator
- 8½" × 11" pieces of light-colored construction paper (nameplates—1 per workshop participant)
- Dark magic markers (1 or 2 per group)
- Pad of giant Post-it notes for groups to record and share ideas
- Index cards (2 colors if possible, but not necessary)
- Scissors (1 or 2 per table)
- Copies of handouts (1 per student)
  - Participant Agenda
  - Case Study: Frustrated
  - Goals and Expectations Worksheet
  - Mentor–Mentee Contract—(give mentors a copy BEFORE the workshop lunch)
  - Three Mentors Worksheet
  - Prioritizing Mentor Roles Worksheet
  - Research Project Design
  - The Next Step in Your Career
  - Letter of Recommendation Worksheet
  - Workshop Evaluation

Getting the Most From Your Research Experience

One-Day Workshop Evaluation

1. What were you expecting from this workshop? Were your expectations met? Please explain.

2. How effectively were the workshop goals met?
   1 = goal not met at all ⇔ 5 = goal met

   Participants will

define, or revisit, their research experience goals and objectives.	1	2	3	4	5
identify and prioritize roles they expect their research mentor to play.	1	2	3	4	5
identify other possible mentors.	1	2	3	4	5
consider the relationships in their research group and how they can most effectively contribute to and learn from the members of this group.	1	2	3	4	5
reflect on the alignment of their research experience goals and expectations with their mentor has for them.	1	2	3	4	5
contribute to defining the elements of a good research project and evaluate their own project in light of these elements.	1	2	3	4	5
consider and begin to plan for the next stage in their career.	1	2	3	4	5

3. How useful did you find the workshop activities?
   1 = not useful at all ⇔ 5 = very useful

Introductory Name Tent Activity	1	2	3	4	5
Establishing Discussion Ground Rules	1	2	3	4	5
Presentation of Workshop Goals	1	2	3	4	5
Case Study: Frustrated	1	2	3	4	5
Goals and Expectations Worksheet	1	2	3	4	5
Three Mentors Activity	1	2	3	4	5
Prioritizing Research Mentor Roles Activity	1	2	3	4	5
Working Lunch with Mentor—Contract	1	2	3	4	5
Elements of a Good Research Project Activity	1	2	3	4	5
Research Project Take Home Worksheets	1	2	3	4	5
The Next Step in Your Career: Prioritizing Activity	1	2	3	4	5
Preparing for the Interview Activity	1	2	3	4	5
Letter of Recommendation Activity	1	2	3	4	5
Overall	**1**	**2**	**3**	**4**	**5**

4. What did you like best about the workshop?

5. How could the workshop be improved?

6. One thing I learned new at this workshop is

7. One thing I am still wondering about after this workshop is

About the Authors

Janet Branchaw is a Faculty Associate at the University of Wisconsin–Madison's Center for Biology Education. She earned her B.S. in Zoology from Iowa State University and her Ph.D. in Physiology from the University of Wisconsin–Madison. After completing postdoctoral training and a lectureship in undergraduate and medical physiology at the University of Wisconsin–Madison's School of Medicine and Public Health, Janet joined the University's Center for Biology Education. Her work at the Center focuses on developing and supporting undergraduate research experiences and preparing a diverse population of undergraduate students for graduate school. She developed and directs two National Science Foundation–funded undergraduate research programs: a 10-week summer Research Experiences for Undergraduates program that hosts students from around the country, and a three-year Undergraduate Research and Mentoring program that prepares students in science, technology, engineering, and math disciplines for interdisciplinary graduate training across the biological sciences. In connection with this work she collaborates with the Delta Program for Research, Teaching and Learning and the Wisconsin Program for Scientific Teaching to train pre-faculty and faculty mentors of undergraduate researchers. In addition to her work in undergraduate research, Janet also teaches introductory biology.

Christine Pfund is the Associate Director of the Delta Program in Research, Teaching, and Learning and the co-Director of the Wisconsin Program for Scientific Teaching at the University of Wisconsin–Madison. She joined these two programs after a post-doctoral research position in Plant Pathology and earning her Ph.D. in Cell and Molecular Biology at University of Wisconsin–Madison. Her work with both the Delta Program and the Program for Scientific Teaching is focused on preparing future faculty to be effective teachers and mentors, as well as successfully integrate their approaches to research with their approaches to teaching and learning. Specifically, she has been integrally involved in developing, implementing, and evaluating a training seminar for research mentors working with undergraduate researchers. She helped develop a manual for facilitators of this seminar, *Entering Mentoring*, and is currently adapting and disseminating it locally and nationally with the support of the Howard Hughes Medical Institute Professors Program and the National Science Foundation. Her current efforts are focused on evaluating the impact of research mentor training on the mentors themselves and the students with whom they work.

Raelyn Rediske is a Research Assistant with the Delta Program in Research, Teaching, and Learning and a graduate student in the School of Education at the University of Wisconsin–Madison. She earned her B.S. in Biological Anthropology from the University of Wisconsin–Madison and her Masters in Education from the Ohio State University in Math, Science and Technology Education. Her thesis research is focused on science communication. She has developed and taught science classes for local outreach programs for the past 10 years and teaches integrated science–language arts classes online for middle school students.

The Scientific Teaching Book Series

The Scientific Teaching Book Series is a collection of practical guides, intended for all science, technology, engineering and mathematics (STEM) faculty who teach undergraduate and graduate students in these disciplines. The purpose of these books is to help faculty become more successful in all aspects of teaching and learning science, including classroom instruction, mentoring students, and professional development. Authored by well-known science educators, the Series provides concise descriptions of best practices and how to implement them in the classroom, the laboratory, or the department. For readers interested in the research results on which these best practices are based, the books also provide a gateway to the key educational literature.

For ongoing information and discussions regarding this and other Scientific Teaching books, please visit: www.whfreeman.com/facultylounge/majorsbio.

Co-editors for the Scientific Teaching Series:

- Sarah Miller, Wisconsin Program for Scientific Teaching, University of Wisconsin, Madison, WI
- William B Wood, University of Colorado Science Education Initiative, Department of Molecular Cell Biology, University of Colorado, Boulder, CO

For further information about the series, please contact:

Susan Winslow, Executive Biology Editor
W.H. Freeman & Co.
41 Madison Ave, New York, NY 10010
swinslow@whfreeman.com